ORDINARY
DIFFERENTIAL EQUATIONS

ORDINARY
DIFFERENTIAL EQUATIONS

I. G. PETROVSKI
Moscow State University

Revised English Edition
Translated and Edited by

Richard A. Silverman

DOVER PUBLICATIONS, INC.
New York

Published in Canada by General Publishing
Company, Ltd., 30 Lesmill Road, Don Mills,
Toronto, Ontario.
Published in the United Kingdom by Constable
and Company, Ltd., 10 Orange Street, London WC 2.

This Dover edition, first published in 1973, is an
unabridged and corrected republication of the work
originally published by Prentice-Hall, Inc. in 1966.

International Standard Book Number: 0-486-61268-6
Library of Congress Catalog Card Number: 72-97098

Manufactured in the United States of America
Dover Publications, Inc.
180 Varick Street
New York, N. Y. 10014

EDITOR'S PREFACE

This is an essentially faithful, but hardly literal, translation of the fifth revised Russian edition (Moscow, 1964) of Professor Petrovski's classic text on ordinary differential equations. In preparing the English edition, I was assisted by Professor E. L. Reiss of New York University, who carefully read the entire typescript and made a number of helpful suggestions. It is my belief that any mathematics book benefits from a critical study by other mathematicians. Thus, as usual, I had no qualms in making various small improvements that occurred to me in the course of working on the book.

In the author's preface to the first and fifth Russian editions, he acknowledges the help of the following colleagues: A. I. Barabanov (whose notes underlie Sections 1–21), V. V. Stepanov, S. A. Galpern, E. M. Landis, A. D. Myshkis (who wrote Chapter 7 on autonomous systems) and O. A. Oleinik (who wrote Section 63 on generalized solutions). Professor Petrovski calls particular attention to the problems, many due to the abovementioned colleagues, which are to be regarded not merely as exercises, but as an inseparable part of the course itself. Finally, in a letter dated October 17, 1965, the author states: "I shall be delighted if my book helps the English-reading mathematical community learn the fundamentals of this theory."

R. A. S.

CONTENTS

PART 2   SYSTEMS OF ORDINARY DIFFERENTIAL EQUATIONS,   Page 87.

4 GENERAL THEORY OF SYSTEMS,   Page 89.

5 LINEAR SYSTEMS: GENERAL THEORY, Page 103.

ORDINARY
DIFFERENTIAL EQUATIONS

Part 1

THE EQUATION $y' = f(x, y)$

1

BASIC CONCEPTS

---

1. Definitions. Examples

By an *ordinary differential equation of order* n we mean a relation of the form

$$F(x, y, y', \ldots, y^{(n)}) = 0,$$

involving a function F of $n + 1$ variables, an independent variable x, an unknown function $y = y(x)$, and the first n derivatives $y', \ldots, y^{(n)}$ of this function with respect to x. A function $y = \varphi(x)$ with derivatives up to order n is called a *solution* of the differential equation if the substitution

$$y = \varphi(x), \ y' = \varphi'(x), \ldots, \ y^{(n)} = \varphi^{(n)}(x)$$

reduces the differential equation to an identity. Unless the converse is explicitly stated, we shall always assume that *the quantities* $x, y, y', \ldots, y^{(n)}$ *take only (finite) real values and all functions are single-valued*. The word "ordinary" means that the unknown function depends on only one independent variable x. We might also study "partial" differential equations, where the unknown function depends on several variables. However, this book is concerned exclusively with ordinary differential equations, except in the Supplement.

Ordinary differential equations play a key role in science and technology, as illustrated by the following two examples:

Example 1. *Describe the motion of a particle moving along the x-axis whose velocity at time t is given by the continuous function $f(t)$ and whose position at time $t = t_0$ is x_0.*

3

Here we must find the solution of the differential equation

$$\frac{dx}{dt} = f(t)$$

which takes the value x_0 at time t_0. Obviously this solution is just

$$x(t) = x_0 + \int_{t_0}^{t} f(\tau)\, d\tau.$$

Example 2. *Given R_0 grams of radium at time t_0, describe the subsequent process of radioactive decay.*

Let $R = R(t)$ be the amount of radium present at time $t \geqslant t_0$. Since the rate of radioactive decay is always proportional to the amount of radium present, with some constant of proportionality $c > 0$, we want the solution of the differential equation

$$\frac{dR}{dt} = -cR$$

which takes the value R_0 at time t_0. It is easily verified by direct substitution that

$$R = R_0 e^{-c(t-t_0)}$$

is the required solution.

In both examples, the differential equation itself is not enough to uniquely determine the unknown function, i.e., in each case there are many functions satisfying the same differential equation. However, in each case there is a unique solution which takes a given value for a given value of the independent variable. Thinking of the independent variable as time, we refer to such a supplementary condition imposed on the solution as an *initial condition*.

Stated concisely, the basic problem of the theory of differential equations is to *find all solutions of a given differential equation and study their properties.* The process of finding solutions of a differential equation is often called *integration* of the equation.

2. Geometric Interpretation. Direction Fields and Integral Curves

Consider the differential equation

$$y' = f(x, y), \qquad (1.1)$$

where the function $f(x, y)$ is defined on some domain of the xy-plane.[1]

---

[1] By a *domain* is meant a nonempty point set G such that

  1. Every point of G is an *interior point*, i.e., has a neighborhood consisting entirely of points of G;

  2. G is *connected*, i.e., any two points of G can be joined by a polygonal line lying entirely in G. (A *polygonal line* is a curve obtained by joining a finite number of line segments end to end.)

Suppose that through every point (x, y) of G we draw a short line segment with slope $f(x, y)$.[2] This gives a set of directions in G, called the *direction field* of (1.1). Then the problem of solving the differential equation (1.1) has the following geometric interpretation: *Find all graphs $y = \varphi(x)$, $a < x < b$ in G whose tangents have directions belonging to the direction field of* (1.1).[3]

From a geometric point of view, the problem as just stated has two unnatural features:

1. By requiring that the slope of the direction field at the point (x, y) of G be given by $f(x, y)$, we exclude directions parallel to the y-axis;

2. By considering only graphs with equations of the form $y = \varphi(x)$, we exclude curves which are intersected more than once by perpendiculars to the x-axis.

Therefore we now restate the problem in a somewhat more general way. First of all, we allow the direction field to be parallel to the y-axis [equivalently, we allow $f(x, y)$ to become infinite]. Thus, besides the differential equation (1.1), we consider the differential equation

$$\frac{dx}{dy} = f_1(x, y), \qquad (1.1')$$

where

$$f_1(x, y) = \frac{1}{f(x, y)}.$$

At points where f and f_1 are both defined, we can use either (1.1) or (1.1'), but we use (1.1') at points where f becomes infinite. (Note that at least one of the functions f and f_1 is always defined, since $f_1 = 0$ if $f = \infty$, while $f = 0$ if $f_1 = \infty$.) In the second place, we allow general curves in parametric form instead of graphs. This leads to the following generalization of the problem posed above: *Find all curves $x = \lambda(t)$, $y = \mu(t)$, $\alpha < t < \beta$ whose tangents have directions specified by the pair of equations* (1.1) *and* (1.1'). Such curves will be called *integral curves* of the equations (1.1) and (1.1'),[4] or of the direction field specified by (1.1) and (1.1'). Thus the graph of every solution of (1.1) is an integral curve of (1.1), (1.1'), but an integral curve of (1.1), (1.1') need not be the graph of a solution of (1.1).

---

[2] The two directions of a line segment will not be distinguished.

[3] A curve is said to be a *graph* if it can be written in the form $y = \varphi(x)$, $a < x < b$, or $x = \psi(y)$, $c < y < d$. The graph is said to be *smooth* if φ or ψ is continuously differentiable.

[4] We shall often use the singular, referring to *equation* (1.1), (1.1').

Remark 1. From now on, by a curve we shall always mean a *smooth* curve with equations of the form $x = \lambda(t)$, $y = \mu(t)$, $\alpha < t < \beta$.[5] The restriction to smooth curves is equivalent to assuming that our solutions are continuously differentiable, and not just differentiable. This class of solutions is large enough for all our subsequent needs.

Remark 2. Sometimes a direction field is specified on some or all of the boundary of the domain G, as well as in G itself.[6] Then integral curves can pass through boundary points of G, as well as interior points.

Remark 3. Henceforth, when dealing with an equation of the form

$$\frac{dy}{dx} = \frac{M(x, y)}{N(x, y)} = f(x, y), \tag{1.2}$$

we shall not bother to explicitly write down the equation

$$\frac{dx}{dy} = \frac{N(x, y)}{M(x, y)} = f_1(x, y). \tag{1.2'}$$

However, we shall often write (1.2) in the form

$$M\,dx - N\,dy = 0, \tag{1.3}$$

which emphasizes the symmetry in x and y. Equation (1.3) specifies a direction field at every point where both factors are defined and at least one is nonzero.

Example 1. The equation

$$\frac{dy}{dx} = \frac{y}{x} \tag{1.4}$$

defines a direction field everywhere except at the origin. This direction field is

---

[5] A curve $x = \lambda(t)$, $y = \mu(t)$, $\alpha < t < \beta$ is said to be *smooth* if $\lambda(t)$ and $\mu(t)$ are both continuously differentiable and if $\lambda'^2(t) + \mu'^2(t) \neq 0$. (The values $\alpha = -\infty$ and $\beta = +\infty$ are not excluded.) Every point of a smooth curve $x = \lambda(t)$, $y = \mu(t)$, $\alpha < t < \beta$ belongs to an arc which is a smooth graph. In fact, given any t_0 ($\alpha < t_0 < \beta$), at least one of the quantities $\lambda'(t_0)$, $\mu'(t_0)$ is nonzero. For example, suppose $\lambda'(t_0) \neq 0$. Then since $\lambda'(t)$ is continuous, it cannot change sign in some interval $t_0 - \varepsilon < t < t_0 + \varepsilon$, and hence we can uniquely solve the equation $x = \lambda(t)$ for t (why?), obtaining a function $t = \gamma(x)$, say. Substituting into the equation $y = \mu(t)$, we find that $y = \mu[\gamma(x)] \equiv \varphi(x)$, $\lambda(t_0 - \varepsilon) < x < \lambda(t_0 + \varepsilon)$. But this is the equation of a (smooth) graph.

[6] By a *boundary point* of a domain G is meant a point which is a limit point of G but does not belong to G. The set of all boundary points of G is called the *boundary* of G. A domain together with its boundary is called a *closed domain*. The closed domain consisting of G and its boundary is called the *closure* of G, denoted by $\bar{G}$.

shown schematically in Figure 1. Every
direction in the field passes through the
origin. Clearly, the function

$$y = kx \qquad (1.5)$$

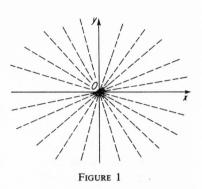

FIGURE 1

is a solution of equation (1.4) for arbitrary finite k. However, the set of all
integral curves of (1.4) is given by the
formula

$$ax + by = 0, \qquad (1.6)$$

where a and b are arbitrary constants,
not both zero. Thus the y-axis is an integral curve of (1.4), but not the graph
of a solution of (1.4).

Since (1.4) does not define a direction field at the origin, we cannot call
(1.5) and (1.6) integral curves at the origin (although they are integral curves
everywhere else). Therefore it would be more accurate to say that the integral
curves are half-lines (rays) emanating from the origin (with the origin itself
excluded), rather than lines passing through the origin.

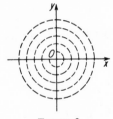

FIGURE 2

Example 2. The equation

$$\frac{dy}{dx} = -\frac{x}{y} \qquad (1.7)$$

defines a direction field everywhere except at the origin.
This direction field is shown schematically in Figure 2.
The directions specified at the point (x, y) by the two
equations (1.4) and (1.7) are orthogonal. Clearly, every
circle with its center at the origin is an integral curve
of equation (1.7). On the other hand, the solutions of (1.4) are the functions

$$y = +\sqrt{R^2 - x^2}, \qquad -R < x < R,$$

$$y = -\sqrt{R^2 - x^2}, \qquad -R < x < R.$$

We conclude this section by introducing some more terminology:

1. For brevity, we shall often say "the solution passes through the point
   (x_0, y_0)" instead of "the graph of the solution passes through the point
   (x_0, y_0)."
2. The function $\varphi(x, C_1, \ldots, C_n)$ involving parameters $C_1, \ldots, C_n$ is
   called the *general solution* of equation (1.1) in a domain G if every
   solution of (1.1) with a graph lying in G is given by φ for a suitable
   choice of $C_1, \ldots, C_n$.

3. The equation $\Phi(x, y) = 0$ of an integral curve of equation (1.1), (1.1')
is called an *integral* of (1.1), (1.1').

4. An equation

$$\Phi(x, y, C_1, \ldots, C_n) = 0 \tag{1.8}$$

involving parameters $C_1, \ldots, C_n$ is called a *complete integral* of equation (1.1), (1.1') in a domain G if every integral curve of (1.1), (1.1')
lying in G is given by (1.8) for a suitable choice of $C_1, \ldots, C_n$.

Thus (1.5) is the general solution of equation (1.4) in the whole xy-plane
minus the y-axis, while (1.6) is the complete integral of (1.4) in the whole
xy-plane minus the origin. In Example 2, the function

$$y = +\sqrt{R^2 - x^2}$$

is the general solution of (1.7) in the half-plane $y > 0$, while the equation

$$x^2 + y^2 = R^2 \tag{1.9}$$

is the complete integral of (1.7) in the whole xy-plane minus the origin. It
will be proved in Sec. 5 that (1.4) has no integral curves other than (1.6),
while (1.7) has no integral curves other than (1.9).

Problem 1. What domain in the plane has no boundary?

Problem 2. Draw the integral curves of the following equations:

a) $\dfrac{dy}{dx} = \dfrac{xy}{|xy|}$; b) $\dfrac{dy}{dx} = \dfrac{|x + y|}{x + y}$; c) $\dfrac{dy}{dx} - = \dfrac{x + |x|}{y + |y|}$;

d) $\dfrac{dy}{dx} = \begin{cases} 0 & \text{for } y \neq x, \\ 1 & \text{for } y = x; \end{cases}$ e) $\dfrac{dy}{dx} = \begin{cases} 1 & \text{for } y \neq x, \\ 0 & \text{for } y = x. \end{cases}$

Find the domain in which each equation defines a direction field.

Problem 3. Given a smooth curve $x = \alpha(t)$, $y = \beta(t)$, $a < t < b$, suppose
$a < a' < b' < b$. Prove that

a) The interval $a' \leqslant t \leqslant b'$ is the union of a finite number of closed intervals
on each of which the curve is a graph of the form $y = \varphi(x)$ or $x = \psi(y)$,
with continuously differentiable φ and ψ;

b) There is a constant $\varepsilon > 0$ such that the arc of the curve corresponding
to the interval $t' \leqslant t \leqslant t' + \varepsilon$ has no self-intersections for every t'
$(a < t' \leqslant b)$;

c) If $l(t_1, t_2)$ is the length of the arc of the curve with end points t_1 and t_2,
then the ratio $l(t_1, t_2)/(t_2 - t_1)$ is bounded and exceeds a positive constant.

Problem 4. Are the following requirements independent:

a) The direction field contains no directions parallel to the y-axis;

b) Every integral curve is the graph of a function of x?

Problem 5. Find the equation of the locus of all points where the solutions of (1.1) have maxima and minima. How about inflection points, in the case where $f(x, y)$ is differentiable?

Problem 6. Verify that substituting curves or even curves with no self-intersections for polygonal lines in the definition of a domain given in footnote 1, p. 4 results in an equivalent definition.

Problem 7. Curves are often specified by equations of the form $f(x, y) = 0$. Suppose the function f is defined and continuously differentiable in the whole plane, where

$$f^2(x, y) + f_x^2(x, y) + f_y^2(x, y) \neq 0, \tag{1.10}$$

and suppose the set $f(x, y) = 0$ is nonempty. Verify that the set consists of a finite or countable number of disjoint curves, where every finite part of the plane can contain only a finite number of curves. Moreover, show that every curve is either closed, or else has both end points at infinity with no self-intersections. What would the set $f(x, y) = 0$ be like if the requirement (1.10) were dropped? How are these assertions changed if the function f is defined on an arbitrary domain G?

2

SOME SIMPLE
DIFFERENTIAL EQUATIONS

---

3. The Equation $y' = f(x)$

We begin by studying the particularly simple differential equation

$$\frac{dy}{dx} = f(x), \tag{2.1}$$

distinguishing two cases:

Case 1. The function $f(x)$ is continuous on (a, b).[1] Then one of the solutions of (2.1) is obviously

$$y(x) = \int_{x_0}^{x} f(\xi) \, d\xi, \tag{2.2}$$

where x_0 and x belong to the interval (a, b), and all other solutions differ from (2.2) by only an additive constant, i.e., the general solution of (2.1) is given by

$$y(x) = \int_{x_0}^{x} f(\xi) \, d\xi + C.$$

Thus subjecting any given integral curve to shifts parallel to the y-axis generates all the other integral curves. Suppose an integral curve is required to pass through a given point (x_0, y_0) of the strip $a < x < b$. Then this uniquely determines the constant C:

$$C = y_0.$$

---

[1] We use parentheses to denote open intervals and brackets to denote closed intervals. Thus (a, b) is the interval $a < x < b$ and $[a, b]$ the interval $a \leqslant x \leqslant b$.

In other words, one and only one integral curve

$$y(x) = y_0 + \int_{x_0}^{x} f(\xi)\, d\xi$$

passes through each point (x_0, y_0) of the strip $a < x < b$.

Case 2. The function $f(x)$ is continuous on (a, b) except at the point c $(a < c < b)$ where $f(x)$ becomes infinite as $x \to c$. Then for $x = c$ we determine the direction field by using the equation

$$\frac{dx}{dy} = \frac{1}{f(x)}. \tag{2.1'}$$

In this case, the direction field becomes steeper and steeper as the line $x = c$ is approached. However, the situation is the same as in Case 1 in the open strips $a < x < c$ and $c < x < b$. Thus if the point (x_0, y_0) belongs to the strip $a < x < c$, there is a unique integral curve

$$y = y_0 + \int_{x_0}^{x} f(\xi)\, d\xi \tag{2.3}$$

lying in the strip and passing through (x_0, y_0). Suppose the integral

$$\int_{x_0}^{x} f(\xi)\, d\xi, \qquad a < x_0 < c \tag{2.4}$$

converges as $x \to c - 0$. Then as $x \to c - 0$ the curve (2.3) approaches a definite point on the line $x = c$, as shown in Figure 3(a). If (2.4) diverges, the curve approaches the line $x = c$ asymptotically as $x \to c - 0$. In the same way, we can study the behavior of the integral curves in the strip $c < x < b$.

Let $f(x) \to +\infty$ as $x \to c \pm 0$, and suppose the integral

$$\int_{x_0}^{x} f(\xi)\, d\xi, \qquad c < x_0 < b \tag{2.4'}$$

converges as $x \to c + 0$ if and only if the integral (2.4) converges as $x \to c - 0$.

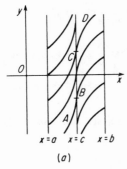 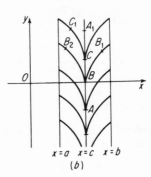

FIGURE 3

Then we get the integral curves shown in Figure 3(a) if the integrals converge, and those shown in Figure 4 if the integrals diverge. If the integrals

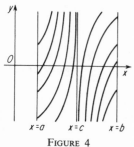

converge, there are infinitely many integral curves of equation (2.1), (2.1′) passing through any given point $A = (x_0, y_0)$ in the strip $a < x < b$. For example, suppose $a < x_0 < c$. Then any curve of the form $ABCD$ [see Figure 3(a)] is an integral curve.

Another possibility is that the integrals (2.4) and (2.4′) both converge as $x \to c \pm 0$, but

FIGURE 4

$$f(x) \to -\infty \quad \text{as} \quad x \to c - 0,$$

$$f(x) \to +\infty \quad \text{as} \quad x \to c + 0.$$

Then the integral curves behave as shown in Figure 3(b). There are infinitely many integral curves AA_1, ABB_1, ACC_1, ... passing through a given point A of the line $x = c$, but there is only one integral curve (say B_1BA) passing through a given point (say B_1) of the strips $a < x < c$ and $c < x < b$, since curves of the form B_1BB_2, with a corner at the point B, are not regarded as integral curves.[2]

If the integrals (2.4) and (2.4′) diverge, there is a unique integral curve passing through every point of the strip $a < x < b$. It is recommended that the reader analyze the case where only one of the integrals (2.4), (2.4′) converges.

Problem 1. Investigate the integral curves of the equation $y' = f(x)$ in the strip $a < x < b$ in the case where $f(x) = 1/\varphi(x)$, $\varphi(c) = 0$ and $\varphi'(c)$ exists ($a < c < b$), assuming that $\varphi(x)$ is continuous for $a < x < b$ and $\varphi(x) \neq 0$ if $x \neq c$.

Problem 2. Examine the behavior of the integral curves of the following equations:

a) $y' = \dfrac{1}{\sqrt[3]{\sin x}}$;   b) $y' = \dfrac{1}{\sqrt[3]{\sin^2 x}}$;   c) $y' = e^{1/x}$;

d) $y' = \dfrac{1}{1 + e^{1/x}}$;   e) $y' = e^{1/x} \sin \dfrac{1}{x}$;   f) $y' = x^a \sin \dfrac{1}{x}$.

(In the last case, allow a to take various values.) In particular, examine the integral curves as $x \to 0$.

Problem 3. Can the equation $y' = f(x)$ have a solution on the entire real axis if the function $f(x)$ is discontinuous?

---

[2] Recall from Remark 1, p. 6 that integral curves are assumed to be smooth.

4. The Equation $y' = f(y)$

The only difference between the differential equation

$$\frac{dy}{dx} = f(y) \tag{2.5}$$

and the equation (2.1) is that the roles of x and y are reversed. If $f(y)$ is continuous and nonzero for $a < y < b$, we can write (2.5) in the form

$$\frac{dx}{dy} = \frac{1}{f(y)}.$$

Therefore there is a unique integral curve

$$x = x_0 + \int_{y_0}^{y} \frac{d\eta}{f(\eta)}$$

passing through every point (x_0, y_0) of the strip $a < y < b$, and subjecting any given integral curve to shifts parallel to the x-axis generates all the other integral curves.

Next suppose $f(y)$ is continuous on (a, b) except at the point $y = c$ $(a < c < b)$ where $f(y) \to 0$ as $y \to c$. Then there are three cases:

1. If

$$\int_{y_0}^{y} \frac{d\eta}{f(\eta)} \tag{2.6}$$

diverges as $y \to c \pm 0$, there is a unique integral curve passing through each point of the strip $a < y < b$, and the line $y = c$ (itself an integral curve) is the asymptote of all the integral curves.

2. If (2.6) converges as $y \to c \pm 0$ and does not change sign as y passes through $y = c$, there are infinitely many integral curves passing through each point of the strip $a < y < b$.

3. If (2.6) converges as $y \to c \pm 0$ and changes sign as y passes through $y = c$, there are infinitely many integral curves passing through each point of the line $y = c$, but one and only one passing through each point of the strips $a < y < c$ and $c < y < b$.

These results all follow from Sec. 3, and the corresponding figures can be obtained by interchanging the x and y-axes in Figures 3 and 4.

Problem 1. Investigate the integral curves of the equation $y' = f(y)$ in the case where $f'(c)$ exists.

Problem 2. Examine the behavior of the integral curves of the following equations:

$$\text{a) } \frac{dy}{dx} = |y|^k; \quad \text{b) } \frac{dy}{dx} = \sin y; \quad \text{c) } \frac{dy}{dx} = \tan\frac{1}{y}.$$

Problem 3. Let $f(c) = 0$ and suppose that arbitrarily close to $y = c$, both for $y < c$ and $y > c$, there are values of y such that $f(y) < 0$ and values of y such that $f(y) > 0$. Prove that $y \equiv c$ is the unique solution of $y' = f(y)$ passing through any point $x = x_0, y = c$.

Problem 4. Prove that every solution of the equation $y' = f(y)$ is monotonic.

Problem 5. Let $f(y)$ be continuous for $a < y < b$, and suppose there is a solution $y = \varphi(x)$ of the equation $y' = f(y)$ such that $\varphi(x) \to c$ as $x \to +\infty$ $(a < c < b)$. Prove that $y \equiv c$ is also a solution.

Problem 6. Let $f(y)$ be continuous and positive for $y_0 \leqslant y < \infty$. Find necessary and sufficient conditions for the solutions of the equation $y' = f(y)$ to have asymptotes. In particular, consider the case

$$f(y) \equiv \frac{P(y)}{Q(y)},$$

where P and Q are polynomials.

Problem 7.[3] Give an example of an equation $y' = f(y)$ with a continuous right-hand side whose solutions include two with the following properties:

1. They are defined and increasing for all x;
2. Their graphs intersect in a single point.

Hint. Let the set of zeros of the function $f(y)$ be a *perfect nowhere dense* set.

Problem 8. Give an example of two equations $y' = f_1(y) \geqslant 0$ and $y' = f_2(y) \geqslant 0$ with continuous right-hand sides such that unique solutions exist for each equation, but not necessarily for the equation

$$y' = \max \{f_1(y), f_2(y)\}.$$

Invent variants of this problem.

5. Equations with Separated Variables

By a differential equation with *separated variables* we mean an equation of the form

$$\frac{dy}{dx} = f_1(x)f_2(y). \tag{2.7}$$

THEOREM. *Suppose $f_1(x)$ is continuous for $a < x < b$ and $f_2(y)$ is continuous and nonzero for $c < y < d$. Then there exists a unique integral curve of equation (2.7) passing through each point (x_0, y_0) of the rectangle $Q: a < x < b, c < y < d$.*

---

[3] For students with a background in point set theory.

Proof. First we prove the uniqueness. Suppose $\varphi(x)$ satisfies (2.7) and the condition $\varphi(x_0) = y_0$. Then we have the identity

$$\frac{d\varphi(x)}{dx} = f_1(x)f_2[\varphi(x)],$$

which can be written as

$$\frac{d\varphi(x)}{f_2[\varphi(x)]} = f_1(x)\,dx, \tag{2.8}$$

since $f_2(y) \neq 0$. Integrating both sides of (2.8) between x_0 and x, we obtain

$$\int_{\varphi(x_0)=y_0}^{\varphi(x)} \frac{d\varphi(\xi)}{f_2[\varphi(\xi)]} = \int_{x_0}^{x} f_1(\xi)\,d\xi.$$

Therefore

$$F_2[\varphi(x)] - F_2(y_0) = F_1(x) - F_1(x_0), \tag{2.9}$$

where $F_2(y)$ is any primitive of $1/f_2(y)$ and $F_1(x)$ any primitive of $f_1(x)$. But since $F_2(y)$ is strictly monotonic [since $F_2'(y) = 1/f_2(y) \neq 0$], we can solve (2.9) uniquely for $\varphi(x)$:[4]

$$\varphi(x) = F_2^{-1}[F_2(y_0) + F_1(x) - F_1(x_0)]. \tag{2.10}$$

Thus, assuming that (2.7) has a solution equal to y_0 for $x = x_0$, we find that this solution must have the form (2.10) and hence must be unique, since all the functions in the right-hand side of (2.10) are completely determined by the differential equation and the initial condition.

To prove the existence of a solution of (2.7) equal to y_0 for $x = x_0$, we need only note that the function $\varphi(x)$ defined by (2.10) actually satisfies the differential equation (in a neighborhood of x_0) and the initial condition. In fact, differentiating (2.9) with respect to x, we obtain

$$\frac{dF_2[\varphi(x)]}{d\varphi(x)}\,\varphi'(x) = F_1'(x)$$

or

$$\frac{1}{f_2[\varphi(x)]}\,\varphi'(x) = f_1(x),$$

i.e., $y = \varphi(x)$ satisfies the differential equation. Moreover, the initial condition is satisfied, since

$$\varphi(x_0) = F_2^{-1}[F_2(y_0)] = y_0.$$

This completes the proof.

---

[4] By F_2^{-1} is meant the inverse of the function F_2.

Remark. If $f(y)$ vanishes at some point $y = y_1$, then uniqueness may be lost, depending on whether or not the integral

$$\int_{y_0}^{y} \frac{d\eta}{f_2(\eta)} \tag{2.11}$$

converges as y approaches y_1. If the integral converges, then infinitely many integral curves pass through certain points of the rectangle Q, and they are all tangent to the integral curve $y = y_1$ corresponding to $f_2(y) = 0$. On the other hand, if (2.11) diverges as $y \to y_1 \pm 0$, there is always a unique integral curve passing through a given point (x_0, y_0) of Q.[5] Of course, here we assume that $f_1(x)$ is not identically zero, since otherwise there will always be a unique integral curve passing through each point of Q.

Problem 1. Investigate the behavior of the integral curves of the following equations:

a) $\dfrac{dy}{dx} = \dfrac{\sin x}{\sin y}$;   b) $\dfrac{dy}{dx} = \dfrac{\sin y}{\sin x}$;   c) $\dfrac{dy}{dx} = \sqrt[3]{\dfrac{x}{y}}$;

d) $\dfrac{dy}{dx} = \sqrt[3]{\dfrac{y}{x}}$;   e) $\dfrac{dy}{dx} = \sqrt[3]{\dfrac{\sin x}{\sin y}}$;   f) $\dfrac{dy}{dx} = \sqrt[3]{\dfrac{\sin y}{\sin x}}$.

Problem 2. Consider the equation

$$\frac{dy}{dx} = \frac{f(y)}{\varphi(x)},$$

where the functions $\varphi(x)$ and $f(y)$ are defined and continuous for all nonnegative x and y, and satisfy the conditions $\varphi(0) = f(0) = 0$. Suppose $\varphi(x)f(y) < 0$ in the domain G: $0 < x < \infty$, $0 < y < \infty$. Investigate the behavior of the integral curves as $x \to \infty$ and also as $y \to \infty$, depending on whether or not the following integrals converge:

$$\int_0^1 \frac{dy}{f(y)}, \quad \int_1^\infty \frac{dy}{f(y)}, \quad \int_0^1 \frac{dx}{\varphi(x)}, \quad \int_1^\infty \frac{dx}{\varphi(x)}.$$

On the other hand, suppose $\varphi(x)f(y) > 0$ in G. Show that every integral curve passing through a point of G approaches the origin arbitrarily closely if the curve is continued in the direction of decreasing x (part of the curve may consist of boundary points of G). Find a criterion for the solution to have an asymptote.

Next suppose that in addition to the above requirements, $\varphi'(0) \neq 0, f'(0) \neq 0$ and $\varphi''(t), f''(t)$ are continuous in some interval $0 \leqslant t < \varepsilon$. Prove that

a) Every integral curve approaches the origin along a definite direction;

---

[5] Uniqueness can only be lost if an integral curve $y = \varphi(x)$ is tangent to the integral curve $y = c$ at an interior point (x_1, y_1) of the rectangle Q. But this is impossible if (2.11) diverges as $y \to y_1$, since then the left-hand side of (2.9) becomes infinite as $x \to x_1$, while the right-hand side remains finite.

   b) If $\varphi'(0) \neq f'(0)$, all the integral curves are tangent to one of the coordinate axes (which one?) for $x = 0$;

   c) If $\varphi'(0) = f'(0)$, the integral curves approach the origin along all directions (as lines $y = kz$).

Finally, make a similar analysis of the solutions of the equations

$$\frac{dy}{dx} = \frac{\varphi(x)}{f(y)}, \qquad \frac{dy}{dx} = \varphi(x)f(y)$$

under the same assumptions.

   Problem 3. Let the functions $\varphi(x)$ and $f(y)$ be continuous and positive for $a < x < b, \, c < y < d$, and suppose each function approaches zero or infinity at the end points of the interval on which it is defined. Study all possible configurations of integral curves of the equation $y' = \varphi(x)f(y)$ in the rectangle $a < x < b, c < y < d$. Then use the result to make a detailed investigation of the integral curves of the equation

$$y' = \frac{P(x)Q(y)}{R(x)S(y)},$$

where P, Q, R and S are polynomials.

6. Homogeneous Equations

By a homogeneous differential equation we mean an equation of the form

$$\frac{dy}{dx} = f\left(\frac{y}{x}\right). \qquad (2.12)$$

Suppose the function $f(u)$ is defined in the strip $a < u < b$. Then the function $f(y/x)$ will be defined in the pair of angles consisting of the points (x, y) such that

$$a < \frac{y}{x} < b.$$

The domain consisting of these two angles will be denoted by G.

   THEOREM. *Suppose $f(u)$ is continuous and never equal to u in the interval $a < u < b$. Then there exists a unique interval curve of equation* (2.12) *passing through every point (x_0, y_0) of G.*

   Proof. Setting $y = ux$, we write (2.12) in the form

$$xu' + u = f(u).$$

But then

$$\frac{du}{dx} = \frac{f(u) - u}{x}, \qquad (2.13)$$

which is a differential equation with separated variables, and the proof is completed by invoking the theorem of the preceding section.

It follows from (2.13) that

$$\frac{dx}{x} = \frac{du}{f(u) - u},$$

and hence

$$\ln |x| = \Phi\left(\frac{y}{x}\right) + C, \qquad (2.14)$$

where $\Phi(u)$ is some primitive of the function

$$\frac{1}{f(u) - u}.$$

We see from (2.14) that all the integral curves of a homogeneous differential equation are similar (in the geometric sense), with the origin as the center of similitude. In fact, there is clearly a choice of the constant C_1 such that the substitution $x \to C_1 x$, $y \to C_1 y$ carries the curve

$$\ln |x| = \Phi\left(\frac{y}{x}\right)$$

into any given curve of the family (2.14).

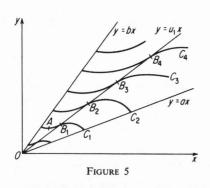

FIGURE 5

The exceptional case $f(u) \equiv u$ has already been considered in Example 1, p. 6. If $f(u) = u$ at separate points $u_1, \ldots, u_n$, then several integral curves may pass through certain points (x_0, y_0) in the domain G, if the integral

$$\int_c^u \frac{d\xi}{f(\xi) - \xi}$$

converges as u approaches one of the numbers $u_1, \ldots, u_n$, say u_1. The behavior of the integral curves in this case is shown schematically in Figure 5. For example, the integral curves AB_1C_1, $AB_1B_2C_2$, $AB_1B_3C_3$, ... pass through the point A and are all tangent to the line $y = u_1 x$.

Problem 1. Investigate the behavior of the integral curves of the following equations:

a) $\dfrac{dy}{dx} = e^{x/y}$;   b) $\dfrac{dy}{dx} = e^{y/x}$;   c) $\dfrac{dy}{dx} = \tan \dfrac{y}{x}$.

Problem 2. Prove that if all the integral curves of a differential equation (2.1) are similar, with the origin as the center of similitude, then the equation is homogeneous.

Problem 3. Suppose $f(u)$ is a continuous function on $0 \leqslant u < u_0$ such that

$$f(0) = 0, \quad f(u) \neq u \qquad (0 < u < u_0).$$

Study all possible configurations of integral curves of equation (2.12) in the sector

$$0 \leqslant \frac{y}{x} \leqslant \frac{u_0}{2} \qquad (x > 0).$$

Then use the result to make a detailed investigation of the integral curves of the equation

$$\frac{dy}{dx} = \frac{P(\sin \varphi, \cos \varphi)}{Q(\sin \varphi, \cos \varphi)},$$

where P and Q are polynomials in two arguments, and φ is the polar angle.

Problem 4. The most general homogeneous equation defined on the whole xy-plane (except at the origin) is of the form

$$\frac{dy}{dx} = F(\varphi), \tag{2.15}$$

where $F(\varphi + 2\pi) \equiv F(\varphi)$. Suppose $F(\varphi)$ can "go to infinity" in a sense like that of Sec. 2, but is otherwise defined and continuous. Under what circumstances can (2.15) be written in the form

$$\frac{dy}{dx} = \frac{f_1(\varphi)}{f_2(\varphi)},$$

where the functions f_1 and f_2 are continuous, finite everywhere and periodic with period 2π, and do not vanish simultaneously?

Problem 5. After transforming to polar coordinates, equation (2.15) of Prob. 4 becomes

$$\frac{d\rho}{d\varphi} = \Phi(\varphi)\rho, \tag{2.16}$$

where the function $\Phi(\varphi)$ has the same properties as the function $F(\varphi)$ of Prob. 4. Equation (2.16) is more convenient for studying solutions. In particular, suppose $\Phi(\varphi)$ is finite everywhere. Describe the behavior of the integral curves when they are continued indefinitely in both directions, as it depends on the sign of the integral

$$\int_0^{2\pi} \Phi(\varphi) \, d\varphi.$$

What happens if the integral vanishes?

Problem 6. Find the most general solution of the equation

$$\frac{dy}{dx} = f\left(\frac{ax + by + c}{a_1 x + b_1 y + c_1}\right),$$

making appropriate assumptions, and describe the behavior of the corresponding integral curves.

7. Linear Equations

By a *linear* differential equation of the first order we mean an equation of the form

$$\frac{dy}{dx} = a(x)y + b(x). \tag{2.17}$$

THEOREM. *Suppose $a(x)$ and $b(x)$ are continuous for $a < x < b$. Then there exists a unique integral curve of equation (2.17) passing through each point (x_0, y_0) of the strip $a < x < b$, $-\infty < y < \infty$ and defined for all x in the interval (a, b).*

Proof.   First we consider the simplest case, where $b(x) \equiv 0$. Then (2.17) reduces to the "homogeneous" linear equation[6]

$$\frac{dy}{dx} = a(x)y, \tag{2.18}$$

which is an equation with separated variables. Recalling the results of Sec. 5 and noting that

$$\int_c^y \frac{d\eta}{\eta}$$

diverges as $y \to 0$, we conclude that (2.18) has a unique solution passing through the point (x_0, y_0). The fact that this solution is given by the formula

$$y(x) = y_0 \exp \left\{ \int_{x_0}^x a(\xi)\, d\xi \right\} \tag{2.19}$$

can be verified by inspection.

We now solve the original "nonhomogeneous" equation (2.17) by the method of *variation of constants*. This is done by writing

$$y(x) = z(x) \exp \left\{ \int_{x_0}^x a(\xi)\, d\xi \right\}, \tag{2.20}$$

where $z(x)$ is now a function of x, instead of a constant as in (2.19). An elementary calculation shows that (2.20) is a solution of (2.17) if and only if $z(x)$ is differentiable and satisfies the equation

$$\frac{dz}{dx} = b(x) \exp \left\{ -\int_{x_0}^x a(\xi)\, d\xi \right\}.$$

---

[6] This use of the word "homogeneous" is not to be confused with the different use of the same word in Sec. 6.

Moreover, $y(x_0) = y_0$ if and only if $z(x_0) = y_0$. Therefore

$$z(x) = y_0 + \int_{x_0}^{x} b(s) \exp\left\{-\int_{x_0}^{s} a(\xi)\, d\xi\right\} ds,$$

and hence

$$y(x) = z(x) \exp\left\{\int_{x_0}^{x} a(\xi)\, d\xi\right\}$$

$$= y_0 \exp\left\{\int_{x_0}^{x} a(\xi)\, d\xi\right\} + \int_{x_0}^{x} b(x) \exp\left\{\int_{s}^{x} a(\xi)\, d\xi\right\} ds$$

is the unique solution of (2.17) which reduces to y_0 for $x = x_0$.

Problem 1. Show that if $n \neq 1$ the equation

$$\frac{dy}{dx} = a(x)y + b(x)y^n$$

(Bernoulli's equation) can be reduced to a linear equation in z by the substitution $z = y^k$, for a suitable choice of k. (For odd n, assume that $y > 0$.) How does one solve Bernoulli's equation if $n = 1$?

Problem 2 (O. A. Oleinik). Let $p(x)$, $q(x)$ and $r(x)$ be continuous functions on $a \leqslant x \leqslant b$ such that

$$p(a) = p(b) = 0, \qquad p(x) > 0 \quad (a < x < b), \qquad q(x) > 0 \quad (a \leqslant x \leqslant b),$$

$$\int_{a}^{a+\varepsilon} \frac{dx}{p(x)} = \int_{b-\varepsilon}^{b} \frac{dx}{p(x)} = +\infty \qquad (0 < \varepsilon < b - a).$$

Prove that all solutions of the equation

$$p(x)\frac{dy}{dx} + q(x)y = r(x)$$

which exist on the interval $a < x < b$ converge to $r(b)/q(b)$ as $x \to b$. Prove that one of these solutions converges to $r(a)/q(a)$ as $x \to a$, while the others converge to $+\infty$ or $-\infty$.

8. Exact Equations. Integrating Factors

As already noted in Sec. 2, it is often convenient to write a differential equation in the form[7]

$$M(x, y)\, dx + N(x, y)\, dy = 0. \tag{2.21}$$

If the left-hand side of (2.21) is an exact differential of some function of x and y, we call (2.21) an *exact* differential equation. Suppose $M(x, y)$ and $N(x, y)$ are continuous and have continuous partial derivatives $\partial M/\partial y$ and $\partial N/\partial x$ on

---

[7] Note that here N has a plus sign, unlike equation (1.3).

some domain G. Then, as we know from calculus, a necessary and sufficient condition for the left-hand side of (2.21) to be an exact differential is that

$$\frac{\partial M}{\partial y} \equiv \frac{\partial N}{\partial x}, \tag{2.22}$$

provided G is *simply connected*.[8]

THEOREM. *Let the functions* $M(x, y)$, $N(x, y)$ *and their partial derivatives* $\partial M/\partial y$, $\partial N/\partial x$ *be continuous on a rectangle* $Q: a < x < b$, $c < y < d$, *and suppose* $N(x, y)$ *is nonvanishing in* Q *and the condition* (2.22) *holds everywhere in* Q. *Then there exists a unique integral curve of equation* (2.21) *passing through each point* (x_0, y_0) *of* Q.[9]

Proof. As already noted, there is a function $z(x, y)$ defined on Q whose exact differential equals the left-hand side of (2.21). This does not involve the assumption that $N \neq 0$. However, since $N \neq 0$, we can write (2.21) in the equivalent form

$$M(x, y) + N(x, y)y' = 0,$$

or

$$\frac{dz[x, y(x)]}{dx} = 0$$

since

$$M \equiv \frac{\partial z}{\partial x}, \qquad N \equiv \frac{\partial z}{\partial y}.$$

Therefore $y(x)$ is a solution of (2.21) if and only if

$$z[x, y(x)] \equiv C = \text{const.} \tag{2.23}$$

This equation can be satisfied by a curve passing through the point (x_0, y_0) only if $C = z(x_0, y_0)$. But if $C = z(x_0, y_0)$, it follows from the implicit function theorem that (2.23) defines a unique curve passing through the point (x_0, y_0), as required. At the same time, we have proved that the desired solution is given by the equation

$$z(x, y) = z(x_0, y_0). \tag{2.24}$$

As we know from calculus, finding the function $z(x, y)$ involves the evaluation of two integrals.

---

[8] A domain G is said to be simply connected if whenever G contains a closed polygonal line L with no self-intersections, it also contains the interior of L (the set of all points inside L).

[9] With these assumptions concerning M and N, it follows from equation (2.21) that dx/dy is nonvanishing in Q. Therefore every integral curve of (2.21) lying in Q must be the graph of a function of x.

Example. Suppose we are given a direction field defined by the equation

$$d\left(\frac{x^2 + y^2}{2}\right) \equiv x\,dx + y\,dy = 0 \qquad (2.25)$$

on the domain between two squares with centers at the origin and sides parallel to the coordinate axes, of lengths 2 and 4 respectively. We cannot apply the above theorem directly to the whole domain G, since $N(x, y) \equiv y$ vanishes on the x-axis. However, we can apply the theorem separately to the four rectangles

$$Q_1: -2 < x < 2, 1 < y < 2,$$
$$Q_2: -2 < x < 2, -2 < y < -1,$$
$$Q_3: \quad 1 < x < 2, -2 < y < 2,$$
$$Q_4: -2 < x < -1, -2 < y < 2,$$

provided that for Q_3 and Q_4 the roles of x and y are reversed in the statement of the theorem. Then it follows from all four results taken together that a unique integral curve of equation (2.25) passes through each point of the domain G. In fact, the curve is just a circle (or a circular arc) with its center at the origin (cf. Example 2, p. 7).

Even if equation (2.22) fails to hold, it is sometimes a comparatively simple matter to transform the differential equation into an exact equation. This is accomplished by using an *integrating factor* $\mu(x, y)$, i.e., a function of x and y such that the left-hand side of (2.21) becomes an exact differential after being multiplied by $\mu(x, y)$. If the functions $M(x, y)$, $N(x, y)$ and $\mu(x, y)$ have continuous partial derivatives, the integrating factor $\mu(x, y)$ must satisfy the condition

$$\frac{\partial(\mu M)}{\partial y} = \frac{\partial(\mu N)}{\partial x},$$

or more explicitly,

$$M\frac{\partial \mu}{\partial y} - N\frac{\partial \mu}{\partial x} = \mu\left(\frac{\partial N}{\partial x} - \frac{\partial M}{\partial y}\right). \qquad (2.26)$$

This is a linear partial differential equation of the first order. To transform the left-hand side of (2.21) into an exact differential, we only need any particular solution of (2.26).[10] However, as we shall see in the Supplement, the problem of finding such a particular solution is in no way simpler than that of finding the general solution of (2.21).

This concludes our discussion of elementary methods for finding solutions of first-order differential equations. For other methods, we refer the reader

---

[10] The trivial solution $\mu \equiv 0$ of (2.26) is obviously of no interest.

to some of the books cited in the Bibliography on p. 227 (especially those by Ince and Kamke). These methods usually have the effect of reducing differential equations to one of the forms considered above.

Problem 1. Is the expression

$$\frac{x\,dx + y\,dy}{x^2 + y^2}$$

the exact differential of some function in the domain considered in the example on p. 23? How about

$$\frac{y\,dx - x\,dy}{x^2 + y^2}\,?$$

Problem 2. Suppose equation (2.21) has continuously differentiable coefficients satisfying the condition (2.22) and defined on a simply connected domain. Show that if (2.21) has a closed integral curve, there is at least one point (x_0, y_0) inside the curve such that

$$M(x_0, y_0) = N(x_0, y_0) = 0.$$

Problem 3. Suppose the condition (2.22) is satisfied in an m-connected domain ($m = 2, 3, \ldots$), i.e., in a domain with m "holes." (Give a precise definition of m-connectedness.) Moreover, suppose the functions $M(x, y)$ and $N(x, y)$ are continuously differentiable and do not vanish simultaneously. Prove that under these conditions the equation of an integral curve passing through a point (x_0, y_0) of the domain can be written in the form (2.24), where however, the function $z(x, y)$ is not single-valued but is defined only to within a term of the form

$$n_1 C_1 + \cdots + n_{m-1} C_{m-1}.$$

Here $C_1, \ldots, C_{m-1}$ are certain completely determined constants (called "periods"), while $n_1, \ldots, n_{m-1}$ are arbitrary integers. Under what conditions will the function $z(x, y)$ be single-valued? What happens if $m = 2$? What is the difference between the case of commensurate periods and the case of incommensurate periods if $m > 2$? What happens if the domain is "infinitely connected?"

Problem 4. Suppose the left-hand side of equation (2.21) becomes the exact differential of a function $z(x, y)$ after being multiplied by a suitable integrating factor $\mu(x, y)$. Show that the function $\mu f[z(x, y)]$, where $f(z)$ is an arbitrary continuous function of z, is also an integrating factor of (2.21). Show that if all relevant functions are continuous and if

$$|M(x_0, y_0)| + |N(x_0, y_0)| > 0, \qquad \mu(x_0, y_0) \neq 0,$$

then any integrating factor of equation (2.21) in some neighborhood of the point (x_0, y_0) can be represented in this form. (However, this last assertion is in general false for the domain as a whole.)

Problem 5. Let $z_1(x, y)$ and $z_2(x, y)$ be continuously differentiable functions on a domain G such that

$$\left| \frac{\partial z_1}{\partial x} \right| + \left| \frac{\partial z_1}{\partial y} \right| > 0, \qquad \begin{vmatrix} \dfrac{\partial z_1}{\partial x} & \dfrac{\partial z_1}{\partial y} \\[2mm] \dfrac{\partial z_2}{\partial x} & \dfrac{\partial z_2}{\partial y} \end{vmatrix} \equiv 0.$$

Prove that z_2 is a function of z_1 in some neighborhood of any point of G (but not necessarily in the whole domain G!).

Problem 6. Find an integrating factor for the linear equation written in the form

$$dy - [a(x)y + b(x)]\, dx = 0.$$

Problem 7. Let $M(x, y)$ and $N(x, y)$ be twice continuously differentiable functions in a rectangle Q, where $N \neq 0$. Prove that a necessary and sufficient condition for the existence in Q of a continuous integrating factor $\mu \neq 0$ [of equation (2.21)] depending only on x is that

$$N \left(\frac{\partial^2 N}{\partial x\, \partial y} - \frac{\partial^2 M}{\partial y^2} \right) \equiv \frac{\partial N}{\partial y} \left(\frac{\partial N}{\partial x} - \frac{\partial M}{\partial y} \right)$$

in Q.

3

GENERAL THEORY

There are relatively few differential equations whose integrals can be found by elementary methods. In fact, as shown by Liouville in 1841, even the simple Riccati equation

$$\frac{dy}{dx} = a_2(x)y^2 + a_1(x)y + a_0(x)$$

cannot be solved by *quadratures*, i.e., by subjecting known functions to a finite number of algebraic operations and integrations (as was done in Secs. 3–8).[1] Therefore a problem of great importance is to develop methods for approximate solution of differential equations which are applicable to large classes of equations. But before trying to find approximate solutions of an equation, we must make sure that the equation has solutions in the first place! The first part of this chapter is devoted to existence questions of this sort. The way in which existence theorems are proved will often suggest methods for approximate determination of solutions (see e.g., Secs. 9, 13, 14 and 17).

9. Euler Lines

Suppose we are given a first-order differential equation

$$y' = f(x, y), \tag{3.1}$$

where $f(x, y)$ is defined on a domain G. As we know, equation (3.1) defines a direction field in G, which in turn has certain integral curves. Choosing any point (x_0, y_0) in G, we draw a line segment through (x_0, y_0) with slope $f(x_0, y_0)$.

---

[1] Concerning the problem of solution by quadratures, see e.g., I. Kaplansky, *An Introduction to Differential Algebra*, Hermann et Cie., Paris (1957).

Then on this segment we choose another point (x_1, y_1) in G, indicated by the number 1 in Figure 6. Through (x_1, y_1) we draw a line segment with slope $f(x_1, y_1)$, and then on this new segment we choose a new point (x_2, y_2) in G, indicated by the number 2 in the figure. Suppose we repeat this construction indefinitely, and let the points be such that

$$x_0 < x_1 < x_2 < \cdots$$

(the construction can also be carried out in the direction of decreasing x). Then the resulting polygonal line will be called an *Euler line*.

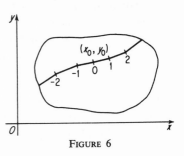

FIGURE 6

It is natural to expect every Euler line passing through (x_0, y_0) to resemble an integral curve passing through (x_0, y_0), if such an integral curve exists and if the segments of the Euler line are sufficiently short. Moreover, we might expect the Euler line to approach an integral curve as the length of its segments goes to zero, and it will be shown later that convergence of this sort actually occurs if $f(x, y)$ is continuous. However, in general there will be no uniqueness, i.e., in general there will be several integral curves passing through the same point (x_0, y_0). In fact, Lavrentev has constructed a differential equation of the form (3.1) such that $f(x, y)$ is continuous and at least two integral curves pass through any given point (x_0, y_0) of G, in every neighborhood of (x_0, y_0).[2] Thus we must assume more than continuity of $f(x, y)$ if only one integral curve is to pass through (x_0, y_0).

In Sec. 11 we shall show that continuity of $f(x, y)$ implies the existence of an integral curve of equation (3.1) passing through any point (x_0, y_0) in G. The proof is due to Peano, and is based on the use of Arzelà's theorem. It is clear that every such integral curve is the *graph* of some *solution* of the differential equation (3.1).

Problem. Let $f(x, y)$ be continuous and bounded on the strip

$$a \leqslant x \leqslant a', \quad -\infty < y < \infty \quad (a < a').$$

Show that the set of all Euler lines of equation (3.1) drawn from a point (a, b) completely fills the region R bounded from above by the curve $y = \varphi_1(x)$, $a \leqslant x \leqslant a'$, from below by the curve $y = \varphi_2(x)$, $a \leqslant x \leqslant a'$ and from the right by the line $x = a'$, where part or all of the curves $y = \varphi_1(x), y = \varphi_2(x)$ may belong to R. Here it is assumed that

   a) $y = \varphi_1(x)$ is concave downward and $y = \varphi_2(x)$ is concave upward;

   b) $\varphi_1(a) = \varphi_2(a) = b$, $\varphi_1'(a) = \varphi_2'(a) = f(a, b)$;

   c) The left and right-hand derivatives of $\varphi_1(x)$ are not exceeded by $f[x, \varphi_1(x)]$, while the left and right-hand derivatives of $\varphi_2(x)$ do not exceed $f[x, \varphi_2(x)]$.

---

[2] M. A. Lavrentev, *Sur une équation differentielle du premier ordre*, Math. Z., **23**, 197 (1925).

10. Arzelà's Theorem

THEOREM. *Let $\{f(x)\}$ be an infinite family of uniformly bounded, equicontinuous functions defined on a finite interval (a, b).*[3] *Then $\{f(x)\}$ contains a uniformly convergent infinite subsequence.*

Proof.[4] The *uniform boundedness* means that there exists a constant M such that

$$|f(x)| < M$$

for every $f(x)$ in the family and every x in (a, b). The *equicontinuity* means that given any $\varepsilon > 0$, there exists a number $\eta = \eta(\varepsilon) > 0$ such that $|x'' - x'| < \eta$ implies

$$|f(x'') - f(x')| < \varepsilon$$

for every $f(x)$ in the family and every x', x'' in (a, b).

Turning to the proof itself, we note that because of the uniform boundedness, the graph of every function in the family $\{f(x)\}$ lies in the rectangle $ABCD$ shown in Figure 7, with sides $2M$ and $b - a$. Consider the infinite sequence

$$\varepsilon_1 = \frac{M}{2^{\alpha+1}}, \quad \varepsilon_2 = \frac{M}{2^{\alpha+2}}, \ldots, \quad \varepsilon_k = \frac{M}{2^{\alpha+k}}, \ldots,$$

where α is any nonnegative integer, and let $\eta_k = \eta(\varepsilon_k)$ be the number associated with ε_k in the definition of equicontinuity. Then partition the rectangle $ABCD$ into subrectangles

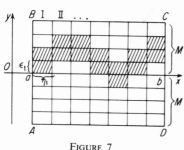

FIGURE 7

by drawing a family of horizontal lines including the x-axis a distance ε_1 apart, and a family of vertical lines including the lines $x = a$ and $x = b$ a distance $\leqslant \eta_1$ apart. Let the corresponding vertical strips be labelled consecutively by Roman numerals I, II, ..., as shown in Figure 7. Since $|x' - x''| < \eta_1$ implies $|f(x') - f(x'')| < \varepsilon_1$, no function in the family $\{f(x)\}$ can have a graph lying in more than two adjacent rectangles belonging to a given strip, in particular to strip I. Therefore there is at least one pair of adjacent rectangles in strip I, e.g., the shaded pair in Figure 7 (where $\alpha = 1$), containing the graphs of infinitely many functions in $\{f(x)\}$. After leaving strip I, the graphs of all the functions in this infinite subfamily of $\{f(x)\}$ can only pass through four rectangles in strip II, while the

---

[3] It does not matter whether the interval is open or closed, i.e., whether or not the interval contains its end points.

[4] Suggested by L. A. Lyusternik.

graph of any given function in the subfamily can only pass through two neighboring rectangles in strip II. As before, there must be two neighboring rectangles in strip II, also shaded in Figure 7, containing the graphs of infinitely many functions in $\{f(x)\}$, each of which falls entirely in the shaded rectangles in strip I as well. Continuing this construction, we finally construct an entire band of width $2\varepsilon_1$, covering the whole interval (a, b) and containing the graphs of infinitely many functions in the family $\{f(x)\}$. This band b_1 is shaded in Figure 7. Let $f_1^*(x)$ be any function in b_1, and let $\{f_1(x)\}$ denote the infinite family consisting of the remaining functions with graphs in b_1.

Next we treat the family $\{f_1(x)\}$ in the same way as the family $\{f(x)\}$, except that we now choose ε_2 instead of ε_1 and η_2 instead of η_1. This leads to a band b_2 of width $2\varepsilon_2$ containing the graphs of infinitely many functions in $\{f_1(x)\}$. Let $f_2^*(x)$ denote any of these functions, and let $\{f_2(x)\}$ denote the family consisting of the remaining functions with graphs in b_2. Repeating this construction indefinitely, we obtain an infinite sequence of functions

$$f_1^*(x), f_2^*(x), \ldots, f_k^*(x), \ldots, \tag{3.2}$$

where the graphs of all the functions starting from $f_k^*(x)$ lie in a band b_k of width $M/2^{k+\alpha-1}$. Therefore the sequence (3.2) is uniformly convergent, and the theorem is proved.

Problem 1. Give examples showing that the uniform boundedness and the equicontinuity are both essential conditions in the statement of Arzelà's theorem, i.e., show that the conclusion of Arzelà's theorem may fail if either condition is dropped.

Problem 2. State and prove Arzelà's theorem for functions of several independent variables.

Problem 3. Does Arzelà's theorem remain true if we allow the interval (a, b) to be infinite?

Problem 4. Show that Arzelà's theorem remains true if instead of uniform boundedness on the whole interval (a, b), we require only that $|f(x_0)| < M$ for every $f(x)$ in the family and some x_0 in (a, b).

II. Peano's Existence Theorem

THEOREM (*Peano*).   *Let $f(x, y)$ be bounded and continuous on a domain G. Then at least one integral curve of the differential equation*

$$\frac{dy}{dx} = f(x, y) \tag{3.3}$$

passes through each point (x_0, y_0) of G.

Proof. Assuming that $|f(x, y)| \leqslant M$, draw the lines with slopes M and $-M$ through the point (x_0, y_0) in G. Then draw the lines $x = a$ and $x = b$ $(a < x_0 < b)$ parallel to the y-axis, which together with the first two lines form two isosceles triangles contained in G with common

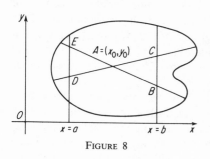

FIGURE 8

vertex (x_0, y_0), as shown in Figure 8. Next, using the method of Sec. 9, construct an infinite sequence of Euler lines

$$L_1, L_2, \ldots, L_k, \ldots$$

passing through the point (x_0, y_0) such that the length of the largest segment of L_k approaches zero as $k \to \infty$. Every polygonal line L_k $(k = 1, 2, \ldots)$ intersects any line parallel to the y-axis only once, and hence is the graph of some continuous function of x, which we denote by $\varphi_k(x)$. The functions

$$\varphi_1(x), \varphi_2(x), \ldots, \varphi_k(x), \ldots$$

have the following properties:

1. They are all defined on the finite closed interval $[a, b]$. In fact, $\varphi_k(x)$ can only fail to be defined on $[a, b]$ if the corresponding Euler line L_k leaves the domain G at some interior point of $[a, b]$. But this is impossible, since the slopes of the segments of L_k are less than M in absolute value, and hence L_k cannot leave the triangles ABC and ADE through the sides BE and DC.

2. They are uniformly bounded on $[a, b]$, since all their graphs are contained in the triangles ABC and ADE.

3. They are equicontinuous on $[a, b]$, since obviously

$$|\varphi_k(x'') - \varphi_k(x')| \leqslant M |x'' - x'|$$

for every $k = 1, 2, \ldots$ and every x', x'' in $[a, b]$.

Thus the sequence $\{\varphi_k(x)\}$ meets all the requirements of Arzelà's theorem, and hence contains a subsequence

$$\varphi^{(1)}(x), \varphi^{(2)}(x), \ldots, \varphi^{(k)}(x), \ldots$$

which converges uniformly on the whole closed interval $[a, b]$ to a limit function

$$\varphi(x) = \lim_{k \to \infty} \varphi^{(k)}(x),$$

where $\varphi(x)$ obviously satisfies the initial condition $\varphi(x_0) = y_0$.

Next we show that the function $\varphi(x)$ satisfies the differential equation (3.3) in the interval (x_0, b) [a similar argument is applicable to the interval (a, x_0)]. Choosing an arbitrary point x' in (x_0, b), we shall prove that

$$\left| \frac{\varphi(x'') - \varphi(x')}{x'' - x'} - f[x', \varphi(x')] \right| \leqslant \varepsilon \qquad (3.4)$$

for arbitrary $\varepsilon > 0$, provided that $|x'' - x'|$ is sufficiently small. To establish (3.4), it is sufficient to prove that

$$\left| \frac{\varphi^{(k)}(x'') - \varphi^{(k)}(x')}{x'' - x'} - f[x', \varphi(x')] \right| < \varepsilon \qquad (3.4')$$

for sufficiently small $|x'' - x'|$ and sufficiently large k, since

$$\varphi^{(k)}(x') \rightarrow \varphi(x'), \qquad \varphi^{(k)}(x'') \rightarrow \varphi(x'')$$

as $k \rightarrow \infty$. Since $f(x, y)$ is continuous on the domain G, given any $\varepsilon > 0$, there is an $\eta > 0$ such that

$$f(x', y') - \varepsilon < f(x, y) < f(x', y') + \varepsilon$$

where $y' = \varphi(x')$,[5] provided that

$$|x - x'| < 2\eta, \qquad |y - y'| < 4M\eta.$$

For sufficiently small η, the set of points (x, y) in G satisfying these inequalities is a rectangle Q (see Figure 9). Now choose K so large that $k > K$ implies that all the segments of L_k are shorter than η and that

$$|\varphi(x) - \varphi^{(k)}(x)| < M\eta$$

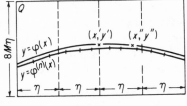

FIGURE 9

for all x in $[a, b]$. Then every Euler line $y = \varphi^{(k)}(x)$ lies entirely in Q if $k > K$ and $|x - x'| < 2\eta$ (why?). Moreover, if $|x'' - x'| < \eta$, we have[6]

$$[f(x', y') - \varepsilon](x'' - x') < \varphi^{(k)}(x'') - \varphi^{(k)}(x') < [f(x', y') + \varepsilon](x'' - x'),$$

which implies (3.4'). This completes the proof.

---

[5] Here it is important not to confuse y' with the derivative dy/dx.

[6] We chose Q to be the rectangle such that $|x - x'| < 2\eta$ instead of simply $|x - x'| < \eta$, in order to be able to estimate the slope of the extreme left-hand segment of an Euler line lying in the strip $S: x' - \eta < x < x' + \eta$. The initial point of this segment may not lie in S, but it will certainly lie in the strip $x' - 2\eta < x < x'$.

Remark 1. The same argument applied to the interval (x_0, b) with $x' = x_0$ shows that the right-hand derivative of $\varphi(x)$ at $x = x_0$, i.e.,

$$\lim_{x'' \to x_0+0} \frac{\varphi(x'') - \varphi(x_0)}{x'' - x_0},$$

equals $f[x_0, \varphi(x_0)]$. Similarly, applying the argument to the interval (x_0, b) with $x' = b$, we see that the left-hand derivative of $\varphi(x)$ at $x = b$, i.e.,

$$\lim_{x'' \to b-0} \frac{\varphi(x'') - \varphi(b)}{x'' - b},$$

equals $f[b, \varphi(b)]$. Similar assertions hold for the end points of the interval (a, x_0).

Remark 2. The above construction leads to a function $\varphi(x)$ equal to y_0 for $x = x_0$ which satisfies equation (3.3), but only on the closed interval $[a, b]$. Consider one of the end points of the resulting arc $y = \varphi(x)$, say the right-hand end point P. By construction P lies in G, and hence we can repeat the whole argument, choosing P instead of (x_0, y_0) and constructing Euler lines to the right of P. In this way, we are able to continue the integral curve up to a point P_1, say. Again we can draw Euler lines through P_1, extending the integral curve further to the right, and so on. This construction leads to integral curves which come arbitrarily close to the boundary of G. In fact, the resulting "chain" of isosceles triangles of the type ABC and ADE must "ultimately shrink to zero," since otherwise the chain would contain infinitely many triangles with lateral side length greater than some number $\varepsilon > 0$. But then we could construct an integral curve covering an arbitrarily large interval of the x-axis, contrary to the assumption that G is bounded. This, together with the fact that sides of triangles like ABC and ADE can come arbitrarily close to the boundary of G, proves our assertion, except for details (see Problem 2).

Remark 3. Peano's theorem remains valid even if $f(x, y)$ is only continuous on G, since every such function is bounded in any domain G' whose closure is contained in G. Using this fact, we can extend Remark 2 to the case of continuous $f(x, y)$. Thus, given a solution of the differential equation (3.3) on the interval $a \leqslant x \leqslant b$, there exists an interval $a' < x < b'$ ($a' < a, b < b'$) and a solution $y = \varphi(x)$ of (3.3) on this interval coinciding with the given solution on $a \leqslant x \leqslant b$ such that one of the following three possibilities occurs:

a) $b' = +\infty$;

b) $|\varphi(x)| \to +\infty$ as $x \to b' - 0$;

c) The distance between the point $(x, \varphi(x))$ and the nearest point of the boundary of G approaches zero as $x \to b' - 0$.

   The situation is the same at the left-hand end point.

Problem 1. Give a detailed proof of Remark 2.

Hint. In constructing the chain of isosceles triangles, let the lateral side length of the "next triangle" equal one half the distance between the "preceding triangle" and the boundary of the domain G. Prove that the resulting sequence of vertices converges to a boundary point of G.

Problem 2. Give a detailed proof of Remark 3.

Problem 3. Suppose the domain G is a strip $a < x < a'$, and let $f(x, y)$ be continuous and bounded on G. It is possible that more than one integral curve of equation (3.3) passes through a given point (x_0, y_0) of G. Prove that there are two integral curves $y = \varphi_1(x)$ and $y = \varphi_2(x)$, the *maximum* and *minimum solutions* (in the sense of Montel) such that

a) $\varphi_1(x_0) = \varphi_2(x_0) = y_0$, $\varphi_2(x) \leqslant \varphi_1(x)$, $a < x < a'$;

b) The entire part of the strip between $y = \varphi_1(x)$ and $y = \varphi_2(x)$ can be completely filled by integral curves passing through (x_0, y_0);

c) There are no integral curves passing through (x_0, y_0) which lie outside this part of the strip.

Problem 4. Prove the existence of solutions of equation (3.3) by the following method, due to Perron: By an *upper function* on the interval (x_0, b) [see Figure 8] is meant any continuously differentiable function $y = \varphi(x)$ whose graph lies entirely in G and which is such that

$$\varphi(x_0) = y_0, \quad \varphi'(x) > f[x, \varphi(x)], \qquad x_0 \leqslant x \leqslant b.$$

Prove that

a) Upper functions exist;

b) The greatest lower bound of all upper functions is an integral curve of equation (3.2) passing through the point (x_0, y_0) [in fact, the maximum solution of Prob. 3].

Similarly, one could define *lower functions* and take their least upper bound. The same thing can be done for (a, x_0).

Problem 5. Let $f(x, y)$ and $F(x, y)$ be two bounded functions defined on a domain G such that $F(x, y) \geqslant f(x, y)$ everywhere on G. Suppose $F(x, y)$ is *upper semicontinuous*, while $f(x, y)$ is *lower semicontinuous*, i.e., suppose that

$$F(x, y) = \limsup_{\substack{t \to x \\ s \to y}} F(t, s),$$

while

$$f(x, y) = \liminf_{\substack{t \to x \\ s \to y}} f(t, s).$$

Prove that there is at least one curve $y = \varphi(x)$ passing through each point (x_0, y_0) of G such that as $\Delta x \to 0$ all the limiting values of the ratio

$$\frac{\varphi(x + \Delta x) - \varphi(x)}{\Delta x}$$

lie between $f(x, y)$ and $F(x, y)$, where x is an arbitrary point of the curve. In general, there are many such curves passing through each point (x_0, y_0), including a maximum and a minimum curve in the sense of Prob. 3.

12. Osgood's Uniqueness Theorem

THEOREM (*Osgood*).   *Suppose the function $f(x, y)$ satisfies the condition*

$$|f(x, y_2) - f(x, y_1)| \leqslant \varphi(|y_2 - y_1|) \tag{3.5}$$

for every pair of points (x, y_1), (x, y_2) in a domain G, where $\varphi(u) > 0$ is a continuous function on $0 < u \leqslant a$ and

$$\lim_{\varepsilon \to 0 + \varepsilon} \int_\varepsilon^a \frac{du}{\varphi(u)} = \infty.$$

Then there is no more than one solution of the equation $y' = f(x, y)$ passing through each point (x_0, y_0) of G.

Proof.   First we note that the functions

$$Ku, \ Ku \, |\ln u|, \ Ku \, |\ln u| \ln |\ln u|, \ Ku \, |\ln u| \ln |\ln u| \cdot \ln \ln |\ln u|, \ldots,$$

where K is a positive constant, all have the properties required of $\varphi(u)$. The most common choice is $\varphi(u) = Ku$. In this case, (3.5) becomes

$$|f(x, y_2) - f(x, y_1)| \leqslant K|y_2 - y_1|,$$

and is known as a *Lipschitz condition*.[7]

Turning to the proof itself, suppose there are two distinct solutions $y_1(x)$ and $y_2(x)$ such that

$$y_1(x_0) = y_2(x_0) = y_0. \tag{3.6}$$

It can be assumed that $x_0 = 0$, since this can always be achieved by replacing x by $x + x_0$. Now let

$$y_2(x) - y_1(x) = z(x).$$

Since $y_2(x) \not\equiv y_1(x)$, there is a point x_1 such that $z(x_1) \neq 0$. We can assume that $z(x_1) > 0$, since the opposite case can be handled merely by setting $z(x)$ equal to $y_1(x) - y_2(x)$ instead of $y_2(x) - y_1(x)$. Similarly, there is no loss of generality in assuming that $x_1 > 0$, since otherwise we need only replace x by $-x$. If $|y_2 - y_1| > 0$, the derivative of z satisfies the inequality

$$\frac{dz}{dx} = \frac{d(y_2 - y_1)}{dx} = f(x, y_2) - f(x, y_1) \leqslant \varphi(|y_2 - y_1|) < 2\varphi(|y_2 - y_1|).$$

$$\tag{3.7}$$

---

[7] Let G be *convex* in y, which means that every line segment AB parallel to the y-axis with end points in G lies entirely in G. Moreover, suppose the partial derivative $\partial f/\partial y$ is bounded everywhere in G. Then f satisfies a Lipschitz condition. In fact, by the law of the mean,

$$|f(x, y_2) - f(x, y_1)| = |f_y[x, y_1 + \theta(y_2 - y_1)]| \, |y_2 - y_1| \leqslant K|y_2 - y_1|,$$

where K is the least upper bound of $|f_y|$ in G.

Next we construct the solution $y(x)$ of the equation

$$\frac{dy}{dx} = 2\varphi(y),$$

satisfying the condition $y(x_1) = z(x_1) = z_1$. Such a solution exists and is unique (see Sec. 4), and its graph approaches the negative x-axis asymptotically without ever intersecting the x-axis (see Figure 10). By
construction, the two curves $z(x)$ and
$y(x)$ intersect at the point (x_1, z_1), and
the inequality

$$z'(x_1) < 2\varphi(z_1) = 2\varphi[y(x_1)] = y'(x_1)$$

immediately implies the existence of
an interval $(x_1 - \varepsilon, x_1), \varepsilon > 0$ on which

$$z(x) > y(x).$$

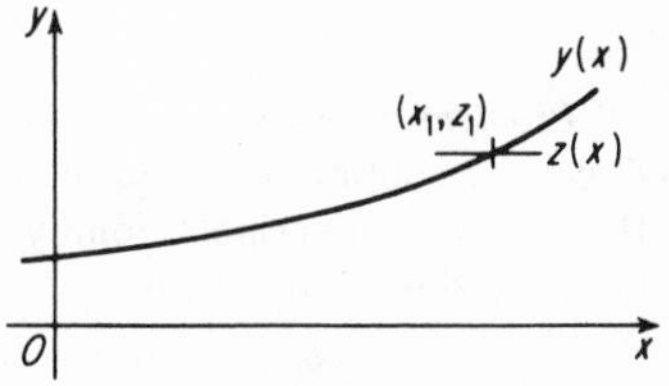

FIGURE 10

But this inequality holds for every ε, if $0 < \varepsilon \leqslant x_1$, since otherwise we immediately arrive at a contradiction by choosing ε to be the largest value compatible with the inequality. In fact, we would then have

$$z'(x_2) \geqslant y'(x_2) = 2\varphi[y(x_2)] = 2\varphi[z(x_2)],$$

since $z(x) > y(x)$ to the right of the point $x_2 = x_1 - \varepsilon$, while on the other hand, since $z(x_2) > 0$, (3.7) implies the contradictory inequality

$$z'(x_2) < 2\varphi[z(x_2)].$$

It follows that

$$z(x) \geqslant y(x) > 0$$

for all x in the interval $[0, x_1]$. In particular, $z(0) = z(x_0) > 0$, contrary to (3.6). Thus there cannot be two distinct solutions of the equation $y' = f(x, y)$, and the proof is complete.

Problem 1. Prove that if $\varphi(0) = 0$, $\varphi'(0) = 0$ and the domain G is convex, then a function $f(x, y)$ satisfying the condition (3.5) must be independent of y. In particular, this is the case if $\varphi(u) = u^p$, $p > 1$.

Problem 2. Prove that if the graph of $\varphi(u)$ is concave downward, then divergence of the integral

$$\int_0^\varepsilon \frac{du}{\varphi(u)}$$

is not only sufficient but also necessary for validity of Osgood's theorem.

Problem 3. Prove that if $\varphi(0) = 0$ and $\varphi'(0)$ exists, then

$$\int_0^\varepsilon \frac{du}{\varphi(u)} = \infty,$$

i.e., the function $\varphi(u)$ satisfies the conditions of Osgood's theorem.

Problem 4. Suppose the function $\varphi(u)$ figuring in Osgood's theorem is chosen in turn to be Ku, $Ku\,|\ln u|$, $Ku\,|\ln u|\,\ln|\ln u|$, ... , and so on. The corresponding progressively weaker restrictions on the function $f(x, y)$ lead to progressively stronger theorems. Prove that there is no "strongest" theorem of this type. In other words, show that if $\varphi(u)$ satisfies the conditions of Osgood's theorem, then there always exists another function $\varphi_1(u)$ satisfying the same conditions such that

$$\lim_{u \to 0} \frac{\varphi_1(u)}{\varphi(u)} = \infty.$$

Problem 5. As we know from analysis, a function $f(x, y)$ defined on a domain G can be continuous in each variable x and y separately, without being continuous in both variables jointly. Prove that if f is continuous in x and satisfies the condition (3.5), where

$$\lim_{u \to 0} \varphi(u) = 0,$$

then f is continuous on G in both x and y. Is this true for functions defined on a square, a disk or a triangle, including its boundary?

Problem 6. Prove that if $\partial f/\partial y$ is continuous on a domain G, then there is a unique solution of $y' = f(x, y)$ passing through each point of G. Is mere existence of this derivative enough?

Problem 7. Let $f(x, y)$ be continuous on a domain G, and let $y(x)$, $z(x)$ and $u(x)$ be continuously differentiable on an interval $x_0 \leqslant x < b$. Moreover, let

$$y(x_0) = z(x_0) = u(x_0) = y_0,$$

where (x_0, y_0) is a point in G, and let

$$y'(x) = f(x, y), \quad z'(x) > f(x, y), \quad u'(x) \geqslant f(x, u)$$

everywhere in $x_0 \leqslant x < b$. Prove that $z(x) > y(x)$ for all $x > x_0$. Assuming in addition that $y(x)$ is the unique solution of $y' = f(x, y)$ on $x_0 \leqslant x < b$, prove that $u(x) \geqslant y(x)$ for all $x > x_0$. Is this still true if the uniqueness requirement on $y(x)$ is dropped? Extend the above assertions to the case where it is only required that $u(x)$ and $z(x)$ be continuous with right-hand (or left-hand) derivatives.

Problem 8. Suppose $f(x, y)$ is continuous on the strip $a \leqslant x \leqslant b$, $-\infty < y < \infty$ and satisfies the estimate $|f(x, y)| \leqslant \varphi(|y|)$, where $\varphi(z) > 0$ is continuous on $0 \leqslant z < \infty$ and

$$\int_N^\infty \frac{dz}{\varphi(z)} = \infty$$

for every $N > 0$. Prove that a solution satisfying arbitrary initial conditions can be continued onto the whole interval $a \leqslant x \leqslant b$. [This is also true if $\varphi(x) \geqslant 0$.] Give an example of a continuous function $\varphi(z) > 0$ such that

$$\int_0^\infty \frac{dz}{\varphi(z)} = \infty$$

but

$$\int_0^\infty \frac{dz}{\varphi(z) + \varepsilon} < \infty$$

for arbitrary $\varepsilon > 0$. What does this imply about continuation of solutions? Is such an example possible if $\varphi(z)$ is monotonic?

Problem 9. Let $y = Y(x)$, $a \leqslant x \leqslant b$ be the maximum solution (in the sense of Prob. 3, Sec. 11) of the differential equation $y' = f(x, y)$ with continuous $f(x, y)$ satisfying the initial conditions $x = x_0$, $Y(x_0) = y_0$ [the curve $y = Y(x)$ lies entirely in G]. Moreover, let $Y_n(x)$ be the maximum solution of $y' = f(x, y)$ satisfying the initial conditions $x = x_n$, $Y_n(x_n) = y_n$, where $y_n \geqslant Y(x_n)$ and

$$\lim_{n \to \infty} x_n = x_0, \qquad \lim_{n \to \infty} y_n = y_0.$$

Show that

a) $Y_n(x)$ exists on the whole interval $[a, b]$ for sufficiently large n;

b) As $n \to \infty$, $Y_n(x)$ converges uniformly on $[a, b]$ to $Y(x)$.

Problem 10. Let $y_n(x)$ be the solution of the equation

$$\frac{dy}{dx} = f(x, y) + \varphi_n(x, y)$$

satisfying the initial condition $y(x_0) = y_0$, where $\varphi_n(x, y) > 0$ and the least upper bound of $\varphi_n(x, y)$ on G converges to zero as $n \to \infty$. Here the functions $f(x, y)$ and $\varphi_n(x, y)$ are continuous, and the point (x_0, y_0) belongs to G. Let $y = Y(x)$ be the same as in the preceding problem. Prove that $y_n(x)$ converges uniformly to $Y(x)$ on the interval $[x_0, b]$ as $n \to \infty$.

Problem 11. Prove a more general uniqueness theorem, replacing (3.5) by the inequality

$$|f(x, y_2) - f(x, y_1)| \leqslant \psi(x)\varphi(|y_2 - y_1|),$$

which holds everywhere except at a finite number of values of x (however, this stipulation can be significantly generalized). Here φ is the same as in Osgood's theorem, while ψ is continuous except possibly at the indicated values of x and

$$\int_a^b \psi(x)\, dx < \infty$$

on any finite interval (a, b). Is the last condition essential?

13. More on Euler Lines

THEOREM.   *Suppose there exists a unique solution $\varphi(x)$ of the equation $y' = f(x, y)$, with continuous $f(x, y)$, and let $L_1, \ldots, L_k, \ldots$ be any sequence of Euler lines drawn from (x_0, y_0) such that $|L_k|$, the length of the largest segment of L_k, approaches zero as $k \to \infty$. Then as $k \to \infty$ the sequence of functions $\varphi_1(x), \ldots, \varphi_k(x), \ldots$, with graphs $L_1, \ldots, L_k, \ldots$*

converges uniformly to $\varphi(x)$ on the closed interval $[a, b]$ figuring in the proof of Peano's existence theorem.

Proof. It is obviously sufficient to prove that given any $\varepsilon > 0$, the sequence $L_1, \ldots, L_k, \ldots$ contains only finitely many Euler lines which do not lie entirely between the curves $y = \varphi(x) - \varepsilon$ and $y = \varphi(x) + \varepsilon$, $a \leqslant x \leqslant b$. Suppose this is not the case. Then there is an infinite sequence of Euler lines $\tilde{L}_1, \ldots, \tilde{L}_k, \ldots$ passing through the point (x_0, y_0), where $|\tilde{L}_k| \to 0$ as $k \to \infty$, none of which lies entirely between the curves $y = \varphi(x) \pm \varepsilon$. But then the same construction as used in the proof of Peano's theorem leads to a sequence of Euler lines uniformly convergent to an integral curve which passes through (x_0, y_0) and is distinct from the curve $y = \varphi(x)$. The fact that this is impossible, because of the assumed uniqueness of $\varphi(x)$, establishes the theorem by contradiction.

Problem 1. Give an example of an equation $y' = f(x, y)$ with continuous $f(x, y)$ and a corresponding sequence of Euler lines $L_1, \ldots, L_k, \ldots$ passing through the point (x_0, y_0) [where $|L_k| \to 0$ as $k \to \infty$] such that the sequence of functions $\varphi_1(x), \ldots, \varphi_k(x), \ldots$ with graphs $L_1, \ldots, L_k, \ldots$ converges at no value of x other than x_0.

Problem 2. Give an example of an equation $y' = f(x, y)$ with continuous $f(x, y)$ such that the limit of a sequence of Euler lines $L_1, \ldots, L_k, \ldots$ beginning at any point (where $|L_k| \to 0$ as $k \to \infty$) always exists and is unique, despite the fact that more than one integral curve passes through certain points of the domain. This possibility stems from the fact that a solution of $y' = f(x, y)$ passing through (x_0, y_0) need not be the limit of a sequence of Euler lines drawn from (x_0, y_0).

14. The Method of Successive Approximations

THEOREM. *Suppose $f(x, y)$ is continuous in x on a domain G and satisfies a Lipschitz condition*

$$|f(x, y_2) - f(x, y_1)| \leqslant K |y_2 - y_1|$$

in y on every closed bounded domain $\bar{G}'$ contained in G (the constant K may depend on the choice of $\bar{G}'$).[8] Then, given any point (x_0, y_0) in G, there

---

[8] It follows from these assumptions that f is continuous in x and y jointly on G. In fact,

$$|f(x_2, y_2) - f(x_1, y_1)| \leqslant |f(x_2, y_2) - f(x_2, y_1)| + |f(x_2, y_1) - f(x_1, y_1)|$$
$$\leqslant K |y_2 - y_1| + |f(x_2, y_1) - f(x_1, y_1)|,$$

where both terms on the right can be made arbitrarily small for (x_2, y_2) sufficiently close to (x_1, y_1) [the second term because of the continuity of f in x].

is a closed interval [a, b] *containing* x_0 *as an interior point on which the differential equation* $y' = f(x, y)$ *has a unique solution* $y(x)$ *equal to* y_0 *for* $x = x_0$.

Proof. First we note that if such a solution $y(x)$ exists, then integration of the identity

$$y'(\xi) = f[\xi, y(\xi)]$$

between x_0 and x gives the *integral equation*[9]

$$y(x) = y_0 + \int_{x_0}^{x} f[\xi, y(\xi)] \, d\xi. \tag{3.8}$$

Thus every solution of the differential equation $y' = f(x, y)$ passing through the point (x_0, y_0) satisfies (3.8). Conversely, every continuous solution $y(x)$ of the integral equation (3.8) satisfies the differential equation and the initial condition $y(x_0) = y_0$.[10] In fact, the initial condition is obviously satisfied by the left-hand side of (3.8), and moreover differentiation of both sides of (3.8) leads at once to the differential equation $y' = f(x, y)$, provided the differentiation is legitimate. But the right-hand side of (3.8) is clearly differentiable for continuous $y(\xi)$ [being an indefinite integral of a continuous function], and hence so is the left-hand side. Therefore, instead of proving that the differential equation $y' = f(x, y)$ has a unique solution equal to y_0 for $x = x_0$ on some closed interval [a, b] ($a < x_0 < b$), we shall now prove that the integral equation (3.8) has a unique continuous solution on [a, b].

Now let $\bar{G}'$ be any closed bounded domain contained in G with (x_0, y_0) as an interior point, and let M be the least upper bound of $|f(x, y)|$ on $\bar{G}'$. As in Figure 8, p. 30, we draw two lines DC and EB with slopes M and $-M$ through the point (x_0, y_0) and then two lines $x = a$ and $x = b$ parallel to the y-axis which together with DC and EB form two isosceles triangles ABC and ADE completely contained in G', with common vertex (x_0, y_0) [later in the proof, we shall assume that a and b are sufficiently close to x_0]. Let $\varphi_0(x)$ be a continuous function on [a, b] whose graph lies in $\bar{G}'$, but which is otherwise completely arbitrary. Substituting $\varphi_0(x)$ into the right-hand side of (3.8), we obtain a new function

$$\varphi_1(x) = y_0 + \int_{x_0}^{x} f[\xi, \varphi_0(\xi)] \, d\xi, \qquad a \leqslant x \leqslant b.$$

It is clear that $\varphi_1(x)$ is itself defined and continuous on [a, b], and in addition satisfies the initial condition $\varphi_1(x_0) = y_0$. Moreover, the

---

[9] So called because the unknown function appears in the integrand. Note that the integrand is a continuous function of ξ [use footnote 8 and the fact that $y(\xi)$ is continuous, being differentiable].

[10] We confine ourselves to continuous solutions of (3.8) only to avoid certain difficulties associated with integration of discontinuous functions.

graph of $\varphi_1(x)$ cannot leave the triangles ABC and ADE, and hence cannot leave $\bar{G}'$, since

$$|f[\xi, \varphi_0(\xi)]| \leqslant M$$

implies

$$|\varphi_1(x) - y_0| \leqslant M |x - x_0|.$$

Next we set

$$\varphi_2(x) = y_0 + \int_{x_0}^x f[\xi, \varphi_1(\xi)] \, d\xi,$$

where the properties of $\varphi_1(x)$ guarantee the existence of the integral on the right. The function $\varphi_2(x)$ is also defined on $[a, b]$, and has the same properties as $\varphi_1(x)$. Similarly, we construct further functions

$$\varphi_3(x) = y_0 + \int_{x_0}^x f[\xi, \varphi_2(\xi)] \, d\xi,$$
$$\cdots\cdots\cdots\cdots\cdots\cdots\cdots\cdots\cdots \tag{3.9}$$
$$\varphi_n(x) = y_0 + \int_{x_0}^x f[\xi, \varphi_{n-1}(\xi)] \, d\xi,$$
$$\cdots\cdots\cdots\cdots\cdots\cdots\cdots\cdots\cdots$$

We now show that the sequence

$$\varphi_0(x), \varphi_1(x), \ldots, \varphi_n(x), \ldots \tag{3.10}$$

constructed in this way converges uniformly on $[a, b]$ to a continuous solution of the integral (3.8). For this reason, the functions (3.10) will be called *successive approximations* to the solution of (3.8).[11] Since $\varphi_n(x)$ can be written in the form

$$\varphi_n(x) = \varphi_0(x) + [\varphi_1(x) - \varphi_0(x)] + \cdots + [\varphi_n(x) - \varphi_{n-1}(x)],$$

it is enough to prove the uniform convergence of the infinite series

$$\varphi_0(x) + [\varphi_1(x) - \varphi_0(x)] + \cdots + [\varphi_n(x) - \varphi_{n-1}(x)] + \cdots. \tag{3.11}$$

With this in mind, we use the Lipschitz condition to estimate $\varphi_{n+1}(x) - \varphi_n(x)$:

$$|\varphi_{n+1}(x) - \varphi_n(x)| = \left| \int_{x_0}^x \{ f[\xi, \varphi_n(\xi)] - f[\xi, \varphi_{n-1}(\xi)] \} \, d\xi \right|$$
$$\leqslant K \left| \int_{x_0}^x |\varphi_n(\xi) - \varphi_{n-1}(\xi)| \, d\xi \right|$$
$$\leqslant K(b - a) \max_{a \leqslant x \leqslant b} |\varphi_n(x) - \varphi_{n-1}(x)|. \tag{3.12}$$

Therefore

$$|\varphi_0| + |\varphi_1 - \varphi_0| + |\varphi_2 - \varphi_1| + |\varphi_3 - \varphi_2| + \cdots$$
$$\leqslant M + M + Mm + Mm^2 + \cdots,$$

----

[11] The method of successive approximations is due to Picard, and can be used to solve a great variety of mathematical problems.

where $K(b - a) = m$, and M_0, M_1 are such that

$$|\varphi_0(x)| \leqslant M_0, \qquad |\varphi_1(x)| \leqslant M_1$$

and hence

$$|\varphi_1(x) - \varphi_0(x)| \leqslant M_0 + M_1 = M.$$

This numerical series converges for $m < 1$. Therefore if we choose $[a, b]$ so small that $K(b - a) = m < 1$, the series (3.11) is uniformly convergent on $[a, b]$ to a continuous function $\varphi(x)$ whose graph lies inside the triangles ABC and ADE. Therefore the integral

$$\int_{x_0}^x f[\xi, \varphi(\xi)] \, d\xi$$

is meaningful. Moreover, since

$$\left| \int_{x_0}^x \{f[\xi, \varphi(\xi)] - f[\xi, \varphi_{n-1}(\xi)]\} \, d\xi \right| \leqslant K \left| \int_{x_0}^x |\varphi(\xi) - \varphi_{n-1}(\xi)| \, d\xi \right|,$$

we can pass to the limit $n \to \infty$ in the relation

$$\varphi_n(x) = y_0 + \int_{x_0}^x f[\xi, \varphi_{n-1}(\xi)] \, d\xi$$

[cf. (3.9)], thereby proving that $\varphi(x)$ satisfies the integral equation (3.8), as required.

We must still prove that (3.8) has no more than one continuous solution on $[a, b]$.[12] Suppose to the contrary that (3.8) has two distinct continuous solutions $\varphi(x)$ and $\tilde{\varphi}(x)$ on $[a, b]$. Then

$$\varphi(x) = y_0 + \int_{x_0}^x f[\xi, \varphi(\xi)] \, d\xi, \qquad \tilde{\varphi}(x) = y_0 + \int_{x_0}^x f[\xi, \tilde{\varphi}(\xi)] \, d\xi,$$

and hence

$$|\varphi(x) - \tilde{\varphi}(x)| = \left| \int_{x_0}^x \{f[\xi, \varphi(\xi)] - f[\xi, \tilde{\varphi}(\xi)]\} \, d\xi \right|$$
$$\leqslant K(b - a) \max_{a \leqslant x \leqslant b} |\varphi(x) - \tilde{\varphi}(x)|.$$

It follows that

$$\max_{a \leqslant x \leqslant b} |\varphi(x) - \tilde{\varphi}(x)| \leqslant K(b - a) \max_{a \leqslant x \leqslant b} |\varphi(x) - \tilde{\varphi}(x)|.$$

But this can only happen if

$$\max_{a \leqslant x \leqslant b} |\varphi(x) - \tilde{\varphi}(x)| = 0,$$

i.e., if $\varphi(x) \equiv \tilde{\varphi}(x)$. Therefore the integral equation (3.8) has a unique continuous solution on $[a, b]$, and the proof is complete.

Remark 1. The sequence (3.10) converges to the same limit function $\varphi(x)$ on $[a, b]$, no matter how we choose the first function $\varphi_0(x)$ of the sequence,

---

[12] Note that such a solution is automatically bounded on $[a, b]$.

provided only that $\varphi_0(x)$ is continuous and has a graph lying entirely in $\bar{G}'$. In fact, as just shown, every such sequence converges to a continuous solution of (3.8), and this solution is unique.

Remark 2. All the remarks made at the end of Sec. 11 remain in force here.

Remark 3. We can often get rid of the condition $K(b - a) < 1$. For example, suppose the domain G contains a strip $c \leqslant x \leqslant d$, $-\infty < y < \infty$ (where c and d are finite), and suppose $f(x, y)$ satisfies a Lipschitz condition

$$|f(x, y_2) - f(x, y_1)| < K|y_2 - y_1|,$$

with the same constant K for all (x, y_1), (x, y_2) in the strip.[13] *Then, if $\varphi_0(x)$ is continuous on $[c, d]$, the successive approximations (3.10), themselves all defined and continuous on $[c, d]$, converge uniformly on $[c, d]$ to a solution of* (3.8). In fact, let

$$M_0 = \max_{c \leqslant x \leqslant d} |\varphi_0(x)|, \qquad M_1 = \max_{c \leqslant x \leqslant d} |\varphi_1(x)|, \qquad M = M_0 + M_1$$

Then, as in (3.12),

$$
\begin{aligned}
|\varphi_2(x) - \varphi_1(x)| &= \left| \int_{x_0}^{x} \{f[\xi, \varphi_1(\xi)] - f[\xi, \varphi_0(\xi)]\}\, d\xi \right| \\
&\leqslant K \left| \int_{x_0}^{x} |\varphi_1(\xi) - \varphi_0(\xi)|\, d\xi \right| \leqslant K \left| \int_{x_0}^{x} M\, d\xi \right| \\
&= \frac{|x - x_0|}{1} MK,
\end{aligned}
$$

$$|\varphi_3(x) - \varphi_2(x)| \leqslant K \left| \int_{x_0}^{x} |\varphi_2(\xi) - \varphi_1(\xi)|\, d\xi \right| \leqslant \frac{|x - x_0|^2}{2} MK^2,$$

and in general

$$|\varphi_{n+1}(x) - \varphi_n(x)| \leqslant \frac{|x - x_0|^n}{n!} MK^n \qquad (c \leqslant x \leqslant d). \tag{3.13}$$

But the series

$$M + M + \frac{|x - x_0|}{1} MK + \frac{|x - x_0|^2}{2} MK^2 + \cdots + \frac{|x - x_0|^n}{n!} MK^n + \cdots$$

converges for all $|x - x_0|$. Therefore the series (3.11) is uniformly convergent on $[c, d]$ (why?). Moreover, it follows from the relation

$$\varphi(x) = \varphi_m(x) + [\varphi_{m+1}(x) - \varphi_m(x)] + [\varphi_{m+2}(x) - \varphi_{m+1}(x)] + \cdots$$

and the estimate (3.13) that

$$|\varphi(x) - \varphi_m(x)| \leqslant MK^m |x - x_0|^m \left[\frac{1}{m!} + K \frac{|x - x_0|}{(m+1)!} + K^2 \frac{|x - x_0|^2}{(m+2)!} + \cdots \right].$$

This allows us to estimate the deviation of the mth approximation from the exact but still unknown solution.

---

[13] This is compatible with $f(x, y)$ being unbounded in the strip.

Remark 4. Sometimes a differential equation is specified in a domain together with some or all of its boundary. Then the situation remains the same if (x_0, y_0) is an interior point. However, if (x_0, y_0) is a boundary point, existence of a solution passing through (x_0, y_0) is far from being guaranteed.[14] However, the same proof as before shows that two distinct solutions satisfying the same initial data cannot exist.

Problem 1. Prove that if $f(x, y)$ is continuously differentiable up to order k, then $\varphi_{k+1}(x)$, $\varphi_{k+2}(x)$, ... are continuously differentiable up to order $k + 1$ and the sequence of jth derivatives of $\varphi_n(x)$, $j = 1, \ldots, k + 1$ converges uniformly to the jth derivative of $\varphi(x)$ on every finite interval where all the $\varphi_n(x)$ exist [and all the graphs $y = \varphi_n(x)$ pass through $\bar{G}'$].

Problem 2. Show that if

$$|\varphi(x) - \varphi_0(x)| \leq c \, |x - x_0|^d \qquad (a \leq x \leq b, a \leq x_0 \leq b, c > 0, d \geq 0),$$

then

$$|\varphi(x) - \varphi_1(x)| \leq c \, |x - x_0|^{d+1} \left[\frac{K}{d + 1} + \frac{K^2}{(d + 1)(d + 2)} + \cdots \right],$$

provided only that the graph of any function $\varphi(x)$ satisfying the given inequality passes through $\bar{G}'$. This estimate allows us to go from the $(n + 1)$th approximation to the exact solution if the $(n + 1)$th approximation differs only slightly from the nth.

Problem 3. Show that the solution of the equation $y' = x^2 + y^2$ on the interval $0 \leq x \leq 1$, subject to the condition $y(0) = 0$, satisfies the inequality

$$|y - (\tfrac{1}{3}x^3 + \tfrac{1}{63}x^7)| \leq 0.0015x^8.$$

Hint. Apply the result of the preceding problem, choosing $\varphi_0(x) = \tfrac{1}{3}x^3$ and studying the given differential equation in the closed domain $0 \leq x \leq 1$, $|y| \leq N$ for suitably chosen $N > 0$.

Problem 4. One way of introducing the concept of a generalized solution of the differential equation $y' = f(x, y)$ in the case of discontinuous $f(x, y)$ is to go over to the integral equation (3.8). Suppose, that $f(x, y)$ is continuous in both x and y everywhere in $\bar{G}$, except possibly at a finite number of "singular" values of x. Suppose further that $f(x, y)$ is bounded and satisfies the inequality

$$|f(x, y_2) - f(x, y_1)| \leq \psi(x) \, |y_2 - y_1|$$

at the "nonsingular" values of x, where $\psi(x)$ is continuous for such x and

$$\int_a^b \psi(x) \, dx < \infty$$

[here (a, b) is any finite interval]. Then define the generalized solution of $y' = f(x, y)$ satisfying the initial condition $y(x_0) = y_0$ to be the continuous solution of the integral equation (3.8). Prove the existence and uniqueness of this solution.

---

[14] Such a solution does exist, for example, if the domain is a strip $x_0 \leq x \leq b$.

Comment. The set of singular values of x may be infinite, but then extra precautions are necessary (this is the "generalized solution in the sense of Carathéodory"[15]).

Problem 5. Choosing $\varphi_0(x) \equiv 1$, examine the behavior of the successive approximations to the solution of $y' = y^2$ on the interval $0 \leqslant x \leqslant 2$, satisfying the initial condition $y(0) = 1$. What happens if the equation is replaced by $y' = y$ or by $y' = -y^2$?

15. The Principle of Contraction Mappings

The method of successive approximations, just used to prove the existence and uniqueness of solutions of differential equations, has applications to many other problems of analysis. This makes it important to find the most general conditions under which the method is applicable. Then instead of setting up the whole method in each special case, we need merely verify that the conditions for applying the method are actually met.

THEOREM 1 (*Principle of contraction mappings*).   *Let* Φ *be a nonempty family of functions* $\varphi(x)$, *each defined on the same set* E, *satisfying the following conditions:*

1) *Every function* φ *is bounded, i.e.,*[16]

$$\sup_{x \in E} |\varphi(x)| = M_\varphi < \infty$$

   (*in general, the bound depends on* φ);

2) *The limit of any uniformly convergent sequence of functions in* Φ *also belongs to* Φ;

3) *There is an operator* A *defined on* Φ *carrying every function* φ *in* Φ *into another function* $A\varphi$ *in* Φ;

4) *The inequality*

$$\sup_{x \in E} |A\varphi(x) - A\tilde{\varphi}(x)| \leqslant m \sup_{x \in E} |\varphi(x) - \tilde{\varphi}(x)|$$

   holds for every pair of functions φ, $\tilde{\varphi}$ *in* Φ *and for some fixed* m *in the interval* $0 \leqslant m < 1$.

Then the equation

$$\varphi = A\varphi \tag{3.14}$$

has one and only one solution belonging to the family Φ.

---

[15] See e.g., G. Sansone, *Equazioni Differenziali nel Campo Reale*, second edition, N. Zanichelli Editore, Bologna (1949), Vol. 2, Chap. 8, Sec. 8.

[16] The symbol $\in$ means "belongs to" or "is an element of." By $\sup_{x \in E}$ is meant the least upper bound with respect to all points $x \in E$.

Proof. Choosing any function φ_0 in the family Φ, we form the function

$$\varphi_1 = A\varphi_0,$$

which we call the "first approximation" to the desired solution of equation (3.14). The function φ_1 belongs to Φ, because of condition 3, and hence we can construct a "second approximation" by setting

$$\varphi_2 = A\varphi_1,$$

where the function φ_2 also belongs to Φ. Continuing this construction indefinitely, we obtain a sequence of functions

$$\varphi_0, \varphi_1, \varphi_2, \ldots, \varphi_k, \ldots, \tag{3.15}$$

where $\varphi_k = A\varphi_{k-1}$ for every $k \geqslant 1$.

We now show that the sequence (3.15) is uniformly convergent on the set E. As in Sec. 14, consider the series

$$\varphi_0 + (\varphi_1 - \varphi_0) + (\varphi_2 - \varphi_1) + \cdots \tag{3.16}$$

If $\sup |\varphi_0| = M_0$ and $\sup |\varphi_1| = M_1$ (recall condition 1), then

$$\sup |\varphi_1 - \varphi_0| \leqslant M_0 + M_1 = M.$$

Applying condition 4, we have

$$\sup |\varphi_{k+1} - \varphi_k| = \sup |A\varphi_k - A\varphi_{k-1}| \leqslant m \sup |\varphi_k - \varphi_{k-1}|.$$

Therefore

$$\sup |\varphi_0| + \sup |\varphi_1 - \varphi_0| + \sup |\varphi_2 - \varphi_1| + \sup |\varphi_3 - \varphi_2| + \cdots$$
$$\leqslant M + M + Mm + Mm^2 + \cdots,$$

and hence the sequence (3.15), whose terms are the partial sums of the series (3.16), converges uniformly on E to some function φ. By condition 2, this limit function φ belongs to the family Φ, and hence it makes sense to write $A\varphi$. Then, by condition 4, we have

$$\sup |A\varphi - A\varphi_{k-1}| \leqslant m \sup |\varphi - \varphi_{k-1}|.$$

It follows that $A\varphi_{k-1}$ converges uniformly to $A\varphi$ as $k \to \infty$, since $\varphi - \varphi_{k-1}$ converges uniformly to zero as $k \to \infty$. Therefore we can pass to the limit as $k \to \infty$ in the relation

$$\varphi_k = A\varphi_{k-1},$$

obtaining

$$\varphi = A\varphi,$$

which agrees with (3.14).

To prove the uniqueness of φ, suppose (3.14) has another solution $\tilde{\varphi}$ in the family Φ. Then

$$\sup |\varphi - \tilde{\varphi}| = \sup |A\varphi - A\tilde{\varphi}| \leqslant m \sup |\varphi - \tilde{\varphi}|,$$

which is possible only if $\sup |\varphi - \tilde{\varphi}| = 0$ or $\varphi \equiv \tilde{\varphi}$, since $m < 1$. Therefore φ is unique, as required.

We now give some examples illustrating the applications of the principle of contraction mappings.

Example 1. First we show how the principle of contraction mappings yields the method of successive approximations. With $[a, b]$ and $\bar{G}'$ the same as in the theorem of Sec. 14, let E and Φ be the interval $[a, b]$ and the family of all continuous functions $\varphi(x)$ defined on $[a, b]$ whose graphs lie in $\bar{G}'$. Then conditions 1 and 2 of Theorem 1 are obviously satisfied. Moreover, let A be the operator carrying φ into

$$A\varphi = y_0 + \int_{x_0}^{x} f[\xi, \varphi(\xi)]\, d\xi.$$

As shown in Sec. 14, this operator satisfies conditions 3 and 4 if the interval $[a, b]$ is sufficiently small. Therefore Theorem 1 implies the existence of a unique continuous solution of the integral equation (3.8), or equivalently the existence of a unique solution $y(x)$ of the differential equation $y' = f(x, y)$ satisfying the initial condition $y(x_0) = y_0$.

Example 2. Consider the integral equation

$$\varphi(x) = f(x) + \lambda \int_{a}^{b} K(x, \xi)\varphi(\xi)\, d\xi, \tag{3.17}$$

where λ is a constant, $f(x)$ is continuous for $a \leqslant x \leqslant b$, and the "kernel" $K(x, \xi)$ is continuous for $a \leqslant x \leqslant b$, $a \leqslant \xi \leqslant b$. Choosing E and Φ to be the closed interval $[a, b]$ and the family of all continuous functions $\varphi(x)$ on $[a, b]$, we immediately verify that conditions 1 and 2 of Theorem 1 are satisfied. Let A be the operator carrying φ into

$$A\varphi = f(x) + \lambda \int_{a}^{b} K(x, \xi)\varphi(\xi)\, d\xi.$$

This operator clearly satisfies condition 3. Moreover,

$$|A\varphi - A\tilde{\varphi}| = \left| \lambda \int_{a}^{b} K(x, \xi)[\varphi(\xi) - \tilde{\varphi}(\xi)]\, d\xi \right|$$

$$\leqslant |\lambda|\, M(b - a) \max_{a \leqslant x \leqslant b} |\varphi(x) - \tilde{\varphi}(x)|,$$

where

$$M = \max_{a \leqslant x, \xi \leqslant b} |K(x, \xi)|,$$

and hence condition 4 is satisfied if $(b - a)|\lambda|M < 1$. Therefore Theorem 1 implies the existence of a unique continuous solution $\varphi(x)$ of the integral equation (3.17) on $a \leqslant x \leqslant b$.

Example 3. Suppose $f(x)$ is defined for all real x and satisfies a Lipschitz condition with constant $K < 1$. Then it follows from the principle of contraction mappings that the equation $x = f(x)$ has a unique solution. In fact, let the set E consist of a single point x_0, and let $\Phi = \{\varphi\}$ be the family of all finite functions defined on E. Then $\{\varphi(x_0)\}$ is just the set of all real numbers, (note that each φ takes only one value). Clearly conditions 1 and 2 of Theorem 1 are satisfied (trivially). Moreover, let A be the operator carrying φ into the function whose value at x_0 is just $f[\varphi(x_0)]$. Since f is defined for all real x, condition 3 is satisfied. To verify condition 4, we merely note that

$$|f(x) - f(\tilde{x})| \leqslant K|x - \tilde{x}| \qquad (K < 1).$$

Finally we use the principle of contraction mappings to prove

THEOREM 2 (*Implicit function theorem*). *Let $f(x, y)$ be defined on the strip $a \leqslant x \leqslant b$, $-\infty < y < \infty$, and suppose $f(x, y)$ is continuous in x and differentiable in y, where*

$$0 < m \leqslant f_y(x, y) \leqslant M < \infty$$

everywhere in the strip. Then the equation

$$f(x, y) = 0 \qquad (3.18)$$

has one and only one continuous solution $y(x)$ on $a \leqslant x \leqslant b$.

Proof. If E is the closed interval $[a, b]$ and Φ the family of all continuous functions on $[a, b]$, then conditions 1 and 2 of Theorem 1 are obviously satisfied. Now let A be the operator carrying φ into

$$A\varphi = \varphi - \frac{1}{M}f(x, \varphi).$$

This operator obviously satisfies condition 3, and moreover condition 4 as well, since

$$|A\varphi - A\tilde{\varphi}| = \left| \varphi - \frac{1}{M}f(x, \varphi) - \left[\tilde{\varphi} - \frac{1}{M}f(x, \tilde{\varphi}) \right] \right|$$

$$= \left| (\varphi - \tilde{\varphi}) - \frac{f_\varphi[x, \tilde{\varphi} + \theta(\varphi - \tilde{\varphi})]}{M}(\varphi - \tilde{\varphi}) \right|$$

$$\leqslant \left(1 - \frac{m}{M} \right)|\varphi - \tilde{\varphi}| \qquad (0 < \theta < 1).$$

It follows from Theorem 1 that the equation

$$\varphi = \varphi - \frac{1}{M} f(x, \varphi),$$

or equivalently the equation (3.18), has a unique continuous solution, as asserted.

Problem 1. Prove that Theorem 1 remains valid if condition 1 is replaced by the weaker condition that the differences between any two functions in Φ be bounded.

Problem 2. Give examples showing that each of the four conditions figuring in Theorem 1 is essential. Also verify that $m = 1$ is not allowed.

16. Geometric Interpretation of the Principle of Contraction Mappings

Suppose we think of the family Φ figuring in the principle of contraction mappings as a set of "points," with the quantity sup $|\varphi - \tilde{\varphi}|$ as the "distance" between two points φ and $\tilde{\varphi}$ in Φ. With this interpretation, condition 2 says that every "limit point" of an infinite sequence of points of the set Φ belongs to Φ, i.e., that the set Φ is closed. Condition 3 means that the operator A carries every point of Φ into another point of Φ, while condition 4 means that if A carries the points φ and $\tilde{\varphi}$ into new points φ^* and $\tilde{\varphi}^*$, then the distance between φ^* and $\tilde{\varphi}^*$ cannot exceed m times the distance between φ and $\tilde{\varphi}$, where $0 \leqslant m < 1$. Finally, finding the solution of equation (3.14) in the class of functions $\Phi = \{\varphi\}$ means finding a point of the set Φ which remains fixed under the action of the operator A.

The fact that such a fixed point exists is geometrically obvious, if it is assumed that the set Φ is bounded. Then the distances between pairs of points of Φ have a least upper bound d, called the "diameter" of the set Φ. Suppose

FIGURE 11

we think of the set Φ as a closed domain, bounded by a curve l (see Figure 11). Then, letting A act upon all the points of Φ, we obtain a new set Φ_1, which is contained in Φ, by condition 3. Moreover, by condition 4, the diameter of Φ_1 is no greater than md. In Figure 11, the set Φ_1 is the region bounded by the curve l_1. Applying the operator A to the points of Φ_1, we obtain a new set Φ_2. Since A carries the point of Φ into points of Φ_1, it certainly carries the points of Φ_1 (a subset of Φ) into points of Φ_1, i.e., Φ_2 is contained in Φ_1. By condition 4, the diameter of Φ_2 cannot exceed m^2d. In Figure 11, the set Φ_2 is the region bounded by the curve l_2.

Continuing this process indefinitely, we obtain an infinite sequence of "nested" closed sets Φ, Φ_1, Φ_2, ..., Φ_n, ..., where the diameter of Φ_n approaches zero as $n \to \infty$. Therefore the intersection of all these sets consists of a single point, which is obviously fixed (invariant) under the action of the operator A. There cannot be two distinct fixed points, since then the distance between the two points would be positive and unaffected by A, contrary to condition 4.

Problem 1. Let F be a closed bounded set in n-dimensional space, and suppose F is mapped into itself by an operator A such that

$$\rho(Ap, Aq) < \rho(p, q), \tag{3.19}$$

where $\rho(p, q)$ denotes the distance between the points p and q. Prove that precisely one point remains fixed. Is this true for sets which are bounded but not closed? How about closed unbounded sets?

Problem 2. Prove that if the inequality (3.19) is replaced by

$$\rho(Ap, Aq) \leqslant m\rho(p, q) \qquad (0 \leqslant m < 1),$$

then there is a unique fixed point for arbitrary closed sets (even if unbounded).

Problem 3. Suppose the inequality (3.19) is replaced by

$$\rho(Ap, Aq) \leqslant \rho(p, q). \tag{3.20}$$

Prove that there is a unique fixed point if F is a line segment or the boundary of an isosceles right triangle, but not if F is a circle or if F is the boundary of an isosceles right triangle and ρ denotes the shortest distance measured along the boundary. Show that there is a unique fixed point if (3.20) holds and if F is any closed bounded convex set (the proof requires the use of a preliminary similarity transformation). Give an example of a closed bounded set and a mapping of this set into itself satisfying (3.20) which has no fixed points and does not reduce to a motion. Can the set be a circle?

Problem 4. Show that if $\varphi(x)$ is a continuous function on $[a, b]$ such that $a \leqslant \varphi(x) \leqslant b$, then there is a point x_0 in $[a, b]$ such that $\varphi(x_0) = x_0$. In other words, show that a continuous transformation of a closed interval into itself has at least one fixed point.

Comment. The analogous theorem is valid for an n-dimensional ball (see Sec. 58).

Problem 5 (O. A. Oleinik). Let $f(x, y)$ be a continuous function defined for all real x and y, such that

   a) $f(x, y)$ satisfies a Lipschitz condition in y;
   b) $f(x + T, y) \equiv f(x, y)$ for some $T > 0$;
   c) $f(x, y_1)f(x, y_2) < 0$ for all x and some y_1, y_2.

Using Prob. 4, prove that the equation $y' = f(x, y)$ has at least one periodic solution, with period T. Apply this result to the equation $y' = a(x)y + b(x)$, where $a(x) \not\equiv 0$ and $b(x)$ are continuous periodic functions, with period T.

17. Cauchy's Theorem

A function $F(x_1, \ldots, x_n)$ is said to be *analytic* in all its arguments in a neighborhood

$$|x_k - x_k^0| < r, \qquad k = 1, \ldots, n \qquad (3.21)$$

of the point $(x_1^0, \ldots, x_n^0)$ if it has a power series expansion in the variables $x_1 - x_1^0, \ldots, x_n - x_n^0$ in the neighborhood (3.21). Here we allow both the function and its arguments to take complex values. By the *derivative* of an analytic function $F(x_1, \ldots, x_n)$ with respect to any argument x_k (in general complex), we mean the function whose power series expansion is obtained by term-by-term differentiation of the power series of $F(x_1, \ldots, x_n)$. As is well known, the differentiated series continues to converge in the neighborhood (3.21).[17] Throughout this section, it does not matter whether the quantities under consideration are real or complex, since the rules for differentiating sums, products, composite functions, etc., are the same in both cases.

THEOREM (*Cauchy's theorem*).[18]   *If $f(x, y)$ is analytic in both x and y in a neighborhood of the point (x_0, y_0), then the differential equation*

$$\frac{dy}{dx} = f(x, y) \qquad (3.22)$$

has a unique solution $y(x)$ satisfying the initial condition

$$y(x_0) = y_0, \qquad (3.23)$$

and this solution is analytic in a neighborhood of x_0.

Proof.   There is no loss of generality in assuming that $x_0 = y_0 = 0$, since otherwise we need only transform to new variables $x^* = x - x_0$, $y^* = y - y_0$. Suppose (3.22) has an analytic solution equal to zero at $x = 0$, with power series

$$y(x) = C_0 + C_1 x + C_2 x^2 + C_3 x^3 + \cdots. \qquad (3.24)$$

Then obviously $C_0 = y(0) = 0$. Substituting (3.24) into (3.22), repeatedly differentiating the result with respect to x, and then setting $x = y = 0$ in each of the equations so obtained, we find the following (infinite)

---

[17] See e.g., T. M. Apostol, *Advanced Calculus*, Addison-Wesley Publishing Co., Inc., Reading, Mass. (1957), Secs. 12–15 and 13–15.

[18] From an historical standpoint, Cauchy's theorem is the first proof of existence and uniqueness of a solution of (3.22), subject to the initial condition (3.23), where $f(x, y)$ is of a rather general form. However, the requirement that $f(x, y)$ be analytic is often artificial, and in fact $f(x, y)$ fails to be analytic in many problems stemming from the applications. Thus a more general theory, like that of Secs. 11, 12 and 14, is indispensable.

system of equations satisfied by the coefficients C_k:

$$C_1 = y'(0) = f(0, 0),$$
$$2C_2 = y''(0) = f_x(0, 0) + f_y(0, 0)y'(0)$$
$$= f_x(0, 0) + f_y(0, 0)C_1,$$
$$3 \cdot 2C_3 = y'''(0) = f_{xx}(0, 0) + 2f_{xy}(0, 0)y'(0) + f_{yy}(0, 0)y'^2(0)$$
$$+ f_y(0, 0)y''(0),$$
$$= f_{xx}(0, 0) + 2f_{xy}(0, 0)C_1 + f_{yy}(0, 0)C_1^2 + f_y(0, 0)2C_2,$$

$$(3.25)$$

. . . .

The coefficients C_k are uniquely determined by (3.25), and hence there can be no more than one solution of (3.22) satisfying the condition (3.23). It is important to note that each C_k can be expressed in terms of sums and products of the coefficients $C_1, \ldots, C_{k-1}$ (with *lower* indices) and the derivatives of $f(x, y)$ at the origin, where the latter are coefficients of the power series expansion

$$f(x, y) = \sum_{k, l=0}^{\infty} a_{kl} x^k y^l \qquad (3.26)$$

(assumed to converge in some neighborhood $|x| < r$, $|y| < r$).

From now on, let (3.24) denote the series *whose coefficients are determined from the system* (3.25). Then the proof will be complete, once we succeed in showing that (3.24) converges in some neighborhood of the point $x = 0$, $y = 0$. In fact, substituting (3.24) into (3.22) and expanding the right-hand side in a (double) power series, we find that identical powers of x have the same coefficients in both sides of the resulting equation.[19] To establish this convergence, we shall consider the auxiliary equation

$$\frac{dz}{dx} = F(x, z), \qquad (3.27)$$

where $F(x, z)$ is analytic in a neighborhood of the origin and has a power series expansion

$$F(x, z) = \sum_{k, l=0}^{\infty} A_{kl} x^k z^l$$

whose coefficients are related to those of (3.26) by the inequalities

$$|a_{kl}| \leqslant |A_{kl}| \qquad (k, l = 0, 1, 2, \ldots).$$

The equation (3.27) is said to *majorize* equation (3.22), and similarly the function $F(x, z)$ is said to *majorize* (or to be a *majorant*) of the function

----

[19] Here we use the theorem on substitution of one power series into another. See e.g., T. M. Apostol, *op. cit.*, p. 414.

$f(x, z)$. If $f(x, y)$ is analytic in the neighborhood $|x| < r$, $|y| < r$, then $f(x, y)$ has a majorant of the form

$$F(x, z) = \frac{M}{\left(1 - \frac{x}{r'}\right)\left(1 - \frac{z}{r'}\right)}, \qquad (|x| < r', |z| < r'),$$

where $0 < r' < r$ and M is a suitable positive constant.[20]

Thus suppose we have found an analytic solution

$$z(x) = C_0^* + C_1^* x + C_2^* x^2 + C_3^* x^3 + \cdots \tag{3.28}$$

of the equation

$$\frac{dz}{dx} = F(x, y) = \frac{M}{\left(1 - \frac{x}{r'}\right)\left(1 - \frac{z}{r'}\right)}$$

satisfying the initial condition $z(0) = 0$ [so that $C_0^* = 0$]. Then the series (3.24) will converge whenever (3.28) does. In fact, each C_k^* is an expression involving sums and products of the coefficients $C_1^*, \ldots, C_{k-1}^*$ and the derivatives of $F(x, y)$ at the origin, and this expression is the same as that for C_k except for the asterisks and the presence of derivatives of $F(x, y)$ instead of derivatives of $f(x, y)$. But $F(x, y)$ majorizes

---

[20] In the case of one variable x, suppose the series

$$f(x) = \sum_{k=0}^{\infty} a_k x^k$$

converges for all x in $(0, r)$. Choosing r' in $(0, r)$, we form the convergent series

$$\sum_{k=0}^{\infty} |a_k| r'^k,$$

with sum M. Then

$$|a_k| r'^k \leqslant M,$$

and hence

$$|a_k| \leqslant \frac{M}{r'^k} \equiv A_k.$$

The function

$$F(x) = \sum_{k=0}^{\infty} A_k x^k = \sum_{k=0}^{\infty} \frac{M}{r'^k} x^k$$

is obviously a majorant of $f(x)$ in $(0, r')$. But clearly

$$F(x) = \sum_{k=0}^{\infty} M \left(\frac{x}{r'}\right)^k = \frac{M}{1 - \frac{x}{r'}}.$$

The generalization to two (or more) variables is immediate.

$f(x, y)$, and hence[21]

$$|C_k| \leqslant C_k^* \qquad (k = 1, 2, \ldots),$$

so that the series (3.24) for $y(x)$ converges by the comparison test. To find $z(x)$, we need only consider real values of x (why?). For real x, we can solve

$$\frac{dz}{dx} = \frac{M}{\left(1 - \dfrac{x}{r'}\right)\left(1 - \dfrac{z}{r'}\right)}$$

by separation of variables, obtaining

$$\int_0^z \left(1 - \frac{\zeta}{r'}\right) d\zeta = \int_0^x \frac{M}{1 - \dfrac{\xi}{r'}} d\xi$$

or

$$z - \frac{z^2}{2r'} = -r'M \ln \left(1 - \frac{x}{r'}\right).$$

This quadratic equation has the solution

$$z(x) = r' - r' \sqrt{1 + 2M \ln \left(1 - \frac{x}{r'}\right)},$$

which vanishes for $x = 0$ and is an analytic function of x for sufficiently small $|x|$, i.e., for

$$|x| < r'(1 - e^{-1/2M})$$

(cf. footnote 19, p. 51). Therefore the formal solution (3.24) converges for sufficiently small $|x|$ to an analytic solution of the differential equation (3.22) satisfying the initial condition (3.23). The proof of Cauchy's theorem is now complete.

COROLLARY.   *If $f(x, y)$ is analytic on a domain G,[22] and if $f(x, y)$ is real for all real x and y, then every real solution of the differential equation $y' = f(x, y)$ whose graph lies in G is analytic.*

Proof.   If $f(x, y)$ is analytic on G, then every point (x_0, y_0) of G has a neighborhood $\mathcal{N}$ in which $f(x, y)$ satisfies a Lipschitz condition in y.

---

[21] For example, according to (3.25),

$$|C_1| = |f(0, 0)| \leqslant F(0, 0) = C_1^*,$$
$$|C_2| \leqslant \tfrac{1}{2}|f(0,0)| + \tfrac{1}{2}|f_y(0, 0)| \, |C_1| \leqslant \tfrac{1}{2}F(0, 0) + \tfrac{1}{2}F_y(0, 0)C_1^* = C_2^*,$$

and so on.

[22] I.e., analytic in a neighborhood of every point of G.

But then, by Osgood's theorem, there can be no more than one solution in $\mathcal{N}$ equal to y_0 for $x = x_0$, and this solution must coincide with the analytic solution constructed above.

Remark. If the function $f(x, y)$ appearing in (3.22) is linear in y, we can find better estimates for the domain of existence of solutions of (3.22) than in the general case. In fact, consider the equation

$$\frac{dy}{dx} = a(x)y + b(x), \tag{3.29}$$

where $a(x)$ and $b(x)$ are analytic functions of x for $|x| < r$. Then (3.29) is majorized by the equation

$$\frac{dz}{dx} = \frac{M}{1 - \dfrac{x}{r'}}\,(z + 1) \qquad (0 < r' < r), \tag{3.30}$$

where the constant M is a common majorant of both $a(x)$ and $b(x)$. But for $|x| < r$, (3.30) has the analytic solution

$$z = \left(1 - \frac{x}{r'}\right)^{-Mr'} - 1,$$

which vanishes for $x = 0$. Therefore, by the same argument as in the general case, the linear equation (3.29) has an analytic solution for $|x| < r'$, equal to zero for $x = 0$.

Problem. Find the first four terms in the expansion of the solution of the equation $y' = e^{xy}$ [$y(0) = 0$] in powers of x. Find upper and lower bounds for the remainder term and for the radius of convergence.

18. Smoothness of Solutions

THEOREM. *If $f(x, y)$ has continuous derivatives with respect to x and y up to order $p \geqslant 0$ (inclusive),*[23] *then every solution of the differential equation*

$$y' = f(x, y) \tag{3.31}$$

has continuous derivatives with respect to x up to order $p + 1$.

Proof. If $y(x)$ is a solution of (3.31), then

$$y'(x) \equiv f[x, y(x)], \tag{3.32}$$

---

[23] If $p = 0$, we regard $f(x, y)$ as continuous.

where in writing (3.32), we presuppose that $y(x)$ is differentiable and hence continuous. Therefore if $f(x, y)$ is continuous in x and y, the right-hand side of (3.32) is a continuous function of x, and hence so is $y'(x)$. Similarly, if $p \geqslant 1$ the right-hand side of (3.32) has a continuous derivative with respect to x, and hence so does $y'(x)$, i.e., $y(x)$ has continuous derivatives up to order 2. Differentiating (3.32) with respect to x, we obtain

$$y''(x) = f_x[x, y(x)] + f_y[x, y(x)]y'(x). \tag{3.33}$$

Assuming that $p \geqslant 2$ and applying the same argument to (3.33) as just applied to (3.32), we find that $y(x)$ has continuous derivatives up to order 3, and so on.

   Problem. Give an example of a differential equation (3.31) where $f(x, y)$ is continuous but not differentiable, all of whose solutions are analytic.

19. Dependence of the Solution of $y' = f(x, y)$ on the Initial Data and $f(x, y)$

Until now, we have considered the problem of finding the solution of a given differential equation $y' = f(x, y)$ passing through a given point (x_0, y_0). In general, of course, changing the point (x_0, y_0) will change the solution, and just *how* the solution changes is a question of great importance in the applications. As pointed out by Hadamard, this question is also fundamental to the whole subject. In fact, suppose the study of some physical problem reduces to finding the solution of a certain differential equation satisfying certain initial conditions. Then these initial conditions are usually found by making experimental measurements. But the ultimate accuracy of such measurements is limited. Therefore a solution $y(x)$ determined by some initial condition $y(x_0) = y_0$ would be of very little interest if slight inaccuracies in the measurement of y_0 (and x_0) could greatly change $y(x)$.

Besides these considerations, it should be pointed out that using a differential equation to study an actual physical process always entails certain idealizations, made with the intention of isolating the most significant features of the problem under discussion. For example, the molecular structure of matter is not taken into account in Example 2, p. 4. Therefore it is in any event only an approximation to describe an actual physical process by a given differential equation and boundary condition, and we would be equally justified in trying to satisfy all differential equations and initial conditions "sufficiently close" to these actually selected, provided of course that these differences do not exceed those allowed by the given idealization of the actual physical process. Thus, in Example 2, p. 4, we could just as well satisfy any other initial condition differing from the given initial condition by less than the mass of a single atom of radium. Therefore the solutions of all "neighboring" differential equations satisfying "neighboring" initial conditions should be "close together," and in fact indistinguishable within the context of the given idealization. Otherwise, using the given differential equation and initial condition to study the physical process in question might not be legitimate.

Having made these observations, we now show that under certain conditions the solution of a differential equation depends continuously on the equation itself and on the initial conditions.

THEOREM.[24] *Let* $f(x, y)$ *be continuous and bounded on a domain* G, *and suppose there is a unique solution of the differential equation*

$$y' = f(x, y) \qquad (3.34)$$

passing through each point (x_0, y_0) *of* G. *Then the solution of* (3.34) *depends continuously on the function* $f(x, y)$ *and on the point* (x_0, y_0) *in the following sense: Suppose the solution* $y_0(x)$ *of* (3.34) *passing through* (x_0, y_0) *is continued onto some interval* $[a, b]$, *where* $a < x_0 < b$. *Let* (x_0^*, y_0^*) *be any point in* G *and* $f^*(x, y)$ *any function continuous on* G. *Then, given any* $\varepsilon > 0$, *there is a* $\delta = \delta(\varepsilon) > 0$ *such that the inequalities*[25]

$$|x_0^* - x_0| < \delta, \qquad |y_0^* - y_0| < \delta, \qquad \sup_{(x,y)\in G} |f^*(x, y) - f(x, y)| < \delta$$

imply that the equation

$$y' = f^*(x, y) \qquad (3.35)$$

has a solution $y_0^*(x)$ *passing through* (x_0^*, y_0^*) *and defined on* $[a, b]$ *which satisfies the inequality*

$$\sup_{a \leqslant x \leqslant b} |y_0^*(x) - y_0(x)| < \varepsilon.$$

Proof. Let (x_k, y_k) be an arbitrary sequence of points in G approaching (x_0, y_0) as $k \to \infty$, and let $\{f_k(x, y)\}$ be an arbitrary sequence of functions continuous on G such that

$$\lim_{k \to \infty} \sup_{(x,y)\in G} |f_k(x, y) - f(x, y)| = 0.$$

Moreover, let $y_k(x)$ be the solution of the differential equation

$$y' = f_k(x, y) \qquad (k = 1, 2, \ldots)$$

satisfying the initial condition $y(x_k) = y_k$. Then we must show that[26]

$$\lim_{k \to \infty} \sup_{a \leqslant x \leqslant b} |y_k(x) - y_0(x)| = 0. \qquad (3.36)$$

---

[24] A related but independent result on the continuous dependence of the solution of (3.34) on the initial data will be proved in Sec. 21 (under stronger assumptions).

[25] Concerning the notation, see footnote 16, p. 44.

[26] This presupposes that $y_k(x)$ can be continued onto the whole interval $[a, b]$, at least for sufficiently large k. This fact will emerge in the course of the proof.

To this end, let M be any number exceeding

$$\sup_{(x,y)\in G} |f(x, y)|,$$

and construct a rectangle of the form

$$|x - x_0| \leqslant c, \qquad |y - y_0| \leqslant Mc$$

lying entirely in G. Then $y_0(x)$ can be defined on the whole interval $|x - x_0| \leqslant c$ and so can $y_k(x)$ for large enough k (why?). Moreover

$$\lim_{k \to \infty} \sup_{|x-x_0|\leqslant c} |y_k(x) - y_0(x)| = 0, \qquad (3.37)$$

since otherwise we could find an infinite sequence of indices $k_1 < k_2 < \cdots$ and an $\varepsilon > 0$ such that

$$\sup_{|x-x_0|\leqslant c} |y_{k_i}(x) - y_0(x)| \geqslant \varepsilon \qquad (3.38)$$

for all $i = 1, 2, \ldots$ But the sequence $\{y_k(x)\}$ is uniformly bounded on $|x - x_0| \leqslant c$ (why?), and moreover equicontinuous on $|x - x_0| \leqslant c$ since for all sufficiently large k we have

$$|y_k'(x)| = |f_k(x, y)| \leqslant M$$

and hence

$$|y_k(x'') - y_k(x')| \leqslant M |x'' - x'|.$$

Therefore, by Arzelà's theorem (see Sec. 10), the sequence $\{y_{k_i}(x)\}$ contains a uniformly convergent subsequence $\tilde{y}_1(x), \tilde{y}_2(x), \ldots$, and by an argument like that given in the proof of Peano's theorem (see Sec. 11), this subsequence converges to a solution of equation (3.34) passing through (x_0, y_0), i.e., to the solution $y_0(x)$ since the latter is unique. But uniform convergence of $\tilde{y}_1(x), \tilde{y}_2(x), \ldots$ to $y_0(x)$ on the interval $|x - x_0| \leqslant c$ contradicts (3.38), thereby establishing (3.37) and proving the theorem if the interval $[a, b]$ figuring in the statement of the theorem is contained in $|x - x_0| \leqslant c$.

If $[a, b]$ is not contained in $|x - x_0| \leqslant c$, we construct another rectangle

$$|x - x_0^{(1)}| \leqslant c_1, \qquad |y - y_0^{(1)}| \leqslant Mc_1$$

lying entirely in G, where

$$x_0^{(1)} = x_0 \pm c, \qquad y_0^{(1)} = y_0(x_0^{(1)}). \qquad (3.39)$$

As just shown, the point $(x_0^{(1)}, y_k(x_0^{(1)}))$ can be made arbitrarily close to $(x_0^{(1)}, y_0^{(1)})$ for sufficiently large k. Therefore, applying the same argument to the solutions $y_0(x), y_k(x)$ and the points $(x_0^{(1)}, y_0^{(1)})$, $(x_0^{(1)}, y_k(x_0^{(1)}))$ in the interval $|x - x_0^{(1)}| \leqslant c_1$ as just applied to the solutions $y_0(x), y_k(x)$ and the points (x_0, y_0), (x_k, y_k) in the interval $|x - x_0| \leqslant c$, we find that

$$\lim_{k \to \infty} \sup_{|x-x_0^{(1)}|\leqslant c_1} |y_k(x) - y_0(x)| = 0.$$

Therefore, taking account of (3.37), we have

$$\lim_{k \to \infty} \sup_{x_0 - c \leqslant x \leqslant x_0 + c + c_1} |y_k(x) - y_0(x)| = 0,$$

if the plus sign was chosen in (3.39). If $[a, b]$ is contained in $[x_0 - c, x_0 + c + c_1]$, the proof is complete. Otherwise we construct further rectangles

$$|x - x_0^{(k)}| \leqslant c_k, \qquad |y - y_0^{(k)}| \leqslant M c_k$$

lying entirely in G, where

$$x_0^{(k)} = x_0^{(k-1)} \pm c_{k-1}, \qquad y_0^{(k)} = y_0(x^{(k)}),$$

until finally, using only a finite number of rectangles (why is this possible and crucial?), we succeed in enlarging the original interval $[x_0 - c, x_0 + c]$ to an interval $[\alpha, \beta]$ containing $[a, b]$ such that

$$\lim_{k \to \infty} \sup_{\alpha \leqslant x \leqslant \beta} |y_k(x) - y_0(x)| = 0.$$

This implies (3.36) and completes the proof.

Remark. Let $f(x, y)$ be continuous and bounded on a domain G, and suppose there is a unique solution $y_0(x)$ of the differential equation $y' = f(x, y)$ passing through *some* point (x_0, y_0) of G. Then *every* solution of (3.35) passing through the point (x_0^*, y_0^*) converges uniformly to $y_0(x)$ as $x_0^* \to x_0$, $y_0^* \to y_0$, $\sup |f^* - f| \to 0$. In fact, this is essentially all that was used to prove the theorem.

Besides proving that the solution depends continuously on the initial data, it is sometimes important to prove the existence of derivatives of the solution with respect to the initial data. Let $y(x, x_0, y_0)$ denote the solution of the differential equation (3.34) satisfying the initial condition $y = y_0$ for $x = x_0$. Suppose we introduce a new function

$$z = y(x, x_0, y_0) - y_0$$

and a new independent variable

$$t = x - x_0,$$

so that $t = 0$, $z = 0$ corresponds to $x = x_0$, $y = y_0$. Then

$$z = y(t + x_0, x_0, y_0) - y_0,$$

and the differential equation (3.34) takes the form

$$\frac{dz}{dt} = f(t + x_0, z + y_0). \tag{3.40}$$

Therefore studying the dependence of the solution of (3.34) on the initial data reduces to studying the dependence of the solution of (3.40) on certain parameters appearing in its right-hand side. This problem was investigated by Poincaré, and will be solved in Sec. 21 by using Hadamard's lemma (proved in Sec. 20).

Problem 1. Prove that if instead of introducing the differential equation (3.34) in the statement of the theorem on p. 56, we simply assume that the function $y_0^*(x)$ satisfies the inequality

$$|y' - f(x, y)| < \delta,$$

then $y_0^*(x)$ can be regarded as piecewise smooth. Also prove that the theorem of Sec. 13 becomes a corollary of the present theorem, if generalized in this way.

Problem 2. Suppose $y(x)$, $z(x)$ and $u(x)$, $a \leqslant x_0 \leqslant b$ are continuous functions satisfying the relations

$$y(x) = y_0 + \int_{x_0}^x f[s, y(s)]\, ds,$$

$$z(x) > y_0 + \int_{x_0}^x f[s, z(s)]\, ds,$$

$$u(x) \geqslant y_0 + \int_{x_0}^x f[s, u(s)]\, ds,$$

where $f(x, y)$ is nondecreasing in y and continuous in a domain containing the graphs of all three functions. Prove that $z(x) > y(x)$ for $x_0 \leqslant x \leqslant b$, and moreover, if the solution of (3.34) satisfying the initial condition $y(x_0) = y_0$ is unique, then $u(x) \geqslant y(x)$ for $x_0 \leqslant x \leqslant b$. [If there is no uniqueness, the right-hand side becomes the minimum solution (as defined in Prob. 3, Sec. 11) of equation (3.34) satisfying the initial condition $y(x_0) = y_0$.] Extend this result to the case of the opposite inequalities. Is it essential that $f(x, y)$ be nondecreasing in y?

Problem 3. Prove the following result, known as *Gronwall's lemma*: If a continuous function $y(x)$, $x_0 \leqslant x \leqslant b$ satisfies the inequality

$$y(x) \leqslant A + \int_{x_0}^x B(s)y(s)\, ds,$$

where the function $B(x) \geqslant 0$ is continuous, then

$$y(x) \leqslant A \exp\left\{ \int_{x_0}^x B(s)\, ds\right\} \qquad (x_0 \leqslant x \leqslant b).$$

Extend this result to the case where $B(x)$ is discontinuous or even unbounded. Under appropriate assumptions, obtain a similar estimate for a function $y(x)$ satisfying the inequality

$$y(x) \leqslant A(x) + \int_{x_0}^x B(s)\, y(s)\, ds.$$

Use Gronwall's lemma to deduce sufficient conditions for uniqueness of the solution of the equation $y' = f(x, y)$.

Problem 4. Find an estimate for a continuous function $y(x)$ satisfying the inequality

$$y(x) \leqslant A + \int_{x_0}^{x} B(s)\varphi[y(s)]\,ds \qquad (x_0 < x),$$

where A is a constant, $\varphi(x) \geqslant 0$ is continuous, $B(x) \geqslant 0$ and

$$\int_{x_0}^{x} B(s)\,ds < \infty.$$

Use this estimate to obtain sufficient conditions for uniqueness of the solution of the equation $y' = f(x, y)$. Compare the result with Osgood's theorem.

Problem 5. Using the result of Prob. 3, find an explicit estimate for the deviation of $y_0^*(x)$ from $y_0(x)$ in the theorem on p. 56, if $f(x, y)$ satisfies a Lipschitz condition in y. Use this result in turn to deduce an explicit estimate for the deviation of an Euler line whose segments have a given maximum length from the exact solution, if $f(x, y)$ satisfies a Lipschitz condition in both variables.

Problem 6. Prove that the continuous dependence proved in the theorem on p. 56 is uniform in x_0^* and y_0^* if the point (x_0^*, y_0^*) belongs to a sufficiently small neighborhood of the curve

$$y = y_0(x) \qquad (a \leqslant x \leqslant b).$$

Problem 7. Consider the equation

$$y' = y(x - h), \tag{3.41}$$

where h is a constant. If $h \geqslant 0$, (3.41) is called a *differential equation with retarded argument*. We are interested in solutions of (3.41) defined for all real x. It is easy to see that knowledge of such a solution on any interval of length h determines the solution everywhere. Prove that if $h > 0$, then given any $A > 0$ and $\varepsilon > 0$, there exists a $\delta > 0$ such that $|y(x)| < \varepsilon$ for $0 \leqslant x \leqslant A$ if $|y(x)| < \delta$ for $-h < x < 0$. However, prove that if $h < 0$, then given any $A > 0$, there exists a sequence of solutions $y_n(x)$ uniformly converging to zero for $-\infty < x < 0$ but such that

$$\lim_{n \to \infty} \sup_{0 < x < A} |y(x)| = \infty.$$

Hint. To prove the last assertion, consider solutions of the form

$$y = e^{\alpha x} \sin \beta x,$$

where α and β are suitably chosen constants.

Problem 8. Given a family of continuous functions defined on an interval $[a, b]$, let A be an arbitrary point of the strip $a \leqslant x \leqslant b$, $-\infty < y < \infty$. Suppose one and only one graph of a function $f_A(x)$ of the family passes through the point A. Prove that $f_{A'}(x)$ converges uniformly to $f_A(x)$ on $[a, b]$ as $A' \to A$ (here A' is a variable point and A a fixed point of the strip).

State an analogous result for a family of functions whose graphs fill a given domain. Roughly speaking, this means that continuous dependence on the initial data follows from existence and uniqueness rather than from the fact that the given functions are solutions of a differential equation.

20. Hadamard's Lemma

LEMMA (*Hadamard*).   *Let G be a domain in $(x_1, \ldots, x_n, z_1, \ldots, z_m)$ space which is convex in $x_1, \ldots, x_n$,[27] and suppose $F(x_1, \ldots, x_n, z_1, \ldots, z_m)$ has continuous derivatives with respect to $x_1, \ldots, x_n$ up to order $p > 0$ (inclusive) on G. Then there are n functions*

$$\Phi_i(x_1, \ldots, x_n, y_1, \ldots, y_n, z_1, \ldots, z_m) \qquad (i = 1, \ldots, n)$$

with continuous derivatives with respect to $x_1, \ldots, x_n, y_1, \ldots, y_n$ up to order $p - 1$ (on G) such that[28]

$$F(y_1, \ldots, y_n, z_1, \ldots, z_m) - F(x_1, \ldots, x_n, z_1, \ldots, z_m)$$
$$= \sum_{i=1}^{n} \Phi_i(x_1, \ldots, x_n, y_1, \ldots, y_n, z_1, \ldots, z_m)(y_i - x_i). \tag{3.42}$$

Proof.[29]   We start from the obvious equality

$$F(y_1, \ldots, y_n, z_1, \ldots, z_m) - F(x_1, \ldots, x_n, z_1, \ldots, z_m)$$
$$= \int_0^1 F_t[x_1 + t(y_1 - x_1), \ldots, x_n + t(y_n - x_n), z_1, \ldots, z_m] \, dt,$$

valid because of the assumed convexity of G in $x_1, \ldots, x_n$. Expressing F_t in terms of derivatives of F with respect to the variables

$$x_1 + t(y_1 - x_1), \ldots, x_n + t(y_n - x_n),$$

and denoting these derivatives by $F_1, \ldots, F_n$, we have

$$F(y_1, \ldots, y_n, z_1, \ldots, z_m) - F(x_1, \ldots, x_n, z_1, \ldots, z_m)$$
$$= \sum_{i=1}^{n} (y_i - x_i) \int_0^1 F_i[x_1 + t(y_1 - x_1), \ldots, x_n + t(y_n - x_n), z_1, \ldots, z_m] \, dt.$$

Choosing the integrals in the right-hand side as the functions Φ_i, we see at once that they have all required properties.

Problem 1. Is the representation of the function F guaranteed by Hadamard's lemma unique?

Problem 2. Prove that if the increment of the function F has the representation (3.42) where the functions Φ_i have continuous derivatives with respect to $x_1, \ldots, x_n, y_1, \ldots, y_n$ up to order $k \geqslant 0$ on G, then the function $F(x_1, \ldots, x_n, z_1, \ldots, z_m)$ has continuous derivatives with respect to $x_1, \ldots, x_n$ up to order $k + 1$ on G. This is to a certain extent the converse of Hadamard's lemma.

---

[27] A domain G in $(x_1, \ldots, x_n, z_1, \ldots, z_m)$ space, i.e., the space of points $(x_1, \ldots, x_n, z_1, \ldots, z_m)$, is said to be *convex* in $x_1, \ldots, x_n$ if whenever two points $(x_1', \ldots, x_n', z_1, \ldots, z_m)$ and $(x_1'', \ldots, x_n'', z_1, \ldots, z_m)$ belong to G, so does the whole line segment joining the two points.

[28] The derivative of order 0 will always mean the function itself.

[29] Due to M. A. Kreines.

21. Dependence of the Solution on Parameters

THEOREM. *Given a differential equation*

$$\frac{dy}{dx} = f(x, y, \mu_1, \ldots, \mu_n), \tag{3.43}$$

let $\bar{G}_{xy}$ be a closed domain in the xy-plane, let P be the open n-dimensional parallelepiped

$$|\mu_1| < \mu_1^0, \ldots, |\mu_n| < \mu_n^0$$

(where the μ_i^0 are positive constants), and let $G = \bar{G}_{xy} \times P$.[30] Suppose $f(x, y, \mu_1, \ldots, \mu_n)$ has bounded continuous derivatives with respect to $y, \mu_1, \ldots, \mu_n$ up to order $p \geqslant 0$ (inclusive) on G,[31] and satisfies a Lipschitz condition

$$|f(x, y_2, \mu_1, \ldots, \mu_n) - f(x, y_1, \mu_1, \ldots, \mu_n)| \leqslant K |y_2 - y_1|$$

in y on G.[32] Then, given any interior point (x_0, y_0) of $\bar{G}_{xy}$, there is a closed interval $a \leqslant x \leqslant b$ $(a < x_0 < b)$ on which the differential equation (3.43) has a unique solution $\varphi(x, \mu_1, \ldots, \mu_n)$ equal to y_0 for $x = x_0$. Moreover, $\varphi(x, \mu_1, \ldots, \mu_n)$ has continuous derivatives with respect to $\mu_1, \ldots, \mu_n$ up to order p on the set $[a, b] \times P$.

Proof. It is sufficient to consider the differential equation

$$\frac{dy}{dx} = f(x, y, \mu), \tag{3.44}$$

whose right-hand side contains only one parameter, since the general case is handled in exactly the same way. Using the method of successive approximations to find a solution of (3.44) equal to y_0 for $x = x_0$, we write

$$\varphi_1(x, \mu) = y_0 + \int_{x_0}^x f[\xi, \varphi_0(\xi), \mu] \, d\xi,$$

$$\varphi_2(x, \mu) = y_0 + \int_{x_0}^x f[\xi, \varphi_1(\xi, \mu), \mu] \, d\xi,$$

and so on, where $\varphi_0(x)$ is continuous on the closed interval $[a, b]$ figuring in the theorem of Sec. 14. It is apparent that every $\varphi_i(x, \mu)$ is continuous in both x and μ on the rectangular region $R: a \leqslant x \leqslant b$, $|\mu| < \mu_0$ in the $x\mu$-plane. Repeating the argument of Sec. 14, we find that the sequence $\{\varphi_i(x, \mu)\}$ converges uniformly in x and μ on R to a limit

---

[30] By $\bar{G}_{xy} \times P$ we mean the *Cartesian product* of $\bar{G}_{xy}$ and P, i.e., the set of all points $(x, y, \mu_1, \ldots, \mu_n)$ such that $(x, y) \in \bar{G}_{xy}$ and $(\mu_1, \ldots, \mu_n) \in P$.

[31] The case $p = 0$ has the same meaning as in footnote 28, p. 61.

[32] Here K is independent of the variables $x, y, \mu_1, \ldots, \mu_n$.

function $\varphi(x, \mu)$, which is the unique solution of (3.44) passing through the point (x_0, y_0). Moreover, $\varphi(x, \mu)$ is clearly continuous in x and μ on R.

Now suppose $p \geqslant 1$, and let $\varphi(x, \mu + \Delta\mu)$ be the unique solution of the equation

$$\frac{dy}{dx} = f(x, y, \mu + \Delta\mu) \tag{3.45}$$

satisfying the initial condition $\varphi(x_0, \mu + \Delta\mu) = y_0$. Substituting $\varphi(x, \mu)$ and $\varphi(x, \mu + \Delta\mu)$ into (3.44) and (3.45), we obtain the identity

$$\frac{d\,\Delta\varphi}{dx} = f[x, \varphi(x, \mu + \Delta\mu), \mu + \Delta\mu] - f[x, \varphi(x, \mu), \mu], \tag{3.46}$$

where $\Delta\varphi$ denotes the increment $\varphi(x, \mu + \Delta\mu) - \varphi(x, \mu)$. According to Hadamard's lemma, the right-hand side of (3.46) is of the form

$$\Phi_1 \Delta\varphi + \Phi_2 \Delta\mu, \tag{3.47}$$

where Φ_1 and Φ_2 are continuous functions of x, $\varphi(x, \mu)$, $\varphi(x, \mu + \Delta\mu)$, μ and $\mu + \Delta\mu$, and hence continuous functions of x and $\Delta\mu$ (if we regard μ as fixed), since $\varphi(x, \mu)$ and $\varphi(x, \mu + \Delta\mu)$ have already been shown to be continuous. Substituting (3.47) into (3.46) and dividing by $\Delta\mu$, we find that $\Delta\varphi/\Delta\mu$ satisfies the linear equation

$$\frac{d}{dx}\left(\frac{\Delta\varphi}{\Delta\mu}\right) = \Phi_1 \frac{\Delta\varphi}{\Delta\mu} + \Phi_2. \tag{3.48}$$

So far, the expression $\Delta\varphi/\Delta\mu$ has only been defined for $\Delta\mu \neq 0$. We now define $\Delta\varphi/\Delta\mu$ for $\Delta\mu = 0$ by requiring that it satisfy (3.48) and vanish for $x = x_0$. Regarding x, $\Delta\mu$ and $\Delta\varphi/\Delta\mu$ as separate variables, we see that the right-hand side of (3.48) is continuous in the variables x and $\Delta\mu$ (figuring in Φ_1 and Φ_2) and has a bounded derivative with respect to $\Delta\varphi/\Delta\mu$. To verify the last assertion, we note that Φ_1 is the integral of certain values of $\partial f/\partial y$ (see Sec. 20) and $\partial f/\partial y$ is bounded. Moreover,

$$\frac{\Delta\varphi}{\Delta\mu} = 0$$

for every $\Delta\mu$ if $x = x_0$. It follows from the first part of the proof that the solution $\Delta\varphi/\Delta\mu$ of equation (3.48) is continuous in $\Delta\mu$ for all sufficiently small values of $|\Delta\mu|$. Therefore $\Delta\varphi/\Delta\mu$ approaches a definite limit as $\Delta\mu \to 0$, and we have proved the existence of the derivative

$$\frac{\partial\varphi(x, \mu)}{\partial\mu}.$$

Moreover, it follows from Hadamard's lemma that

$$\lim_{\Delta\mu\to 0}\Phi_1 = \frac{\partial f}{\partial\varphi}, \qquad \lim_{\Delta\mu\to 0}\Phi_2 = \frac{\partial f}{\partial\mu},$$

and hence $\partial\varphi/\partial\mu$ satisfies the differential equation

$$\frac{d}{dx}\left(\frac{\partial\varphi}{\partial\mu}\right) = \frac{\partial f}{\partial\varphi}\frac{\partial\varphi}{\partial\mu} + \frac{\partial f}{\partial\mu} \tag{3.49}$$

and the initial condition

$$\frac{\partial\varphi}{\partial\mu}\bigg|_{x=x_0} = 0.$$

Therefore, applying the first part of the theorem to (3.49), we find that $\partial\varphi/\partial\mu$ is continuous in x and μ on the region R.

If $p \geqslant 2$, we apply the same argument to equation (3.49) as just applied to equation (3.44), with $\partial\varphi/\partial\mu$ now playing the role of φ. In this way, we find that $\partial^2\varphi/\partial\mu^2$ exists and is continuous in x and μ on R. The theorem in its full generality is proved by repeating this argument (and extending it to equations with several parameters $\mu_1, \ldots, \mu_n$).

COROLLARY. *If $f(x, y)$ has continuous derivatives with respect to x and y up to order $p \geqslant 1$, then the function $y(x, x_0, y_0)$ satisfying the differential equation*

$$y' = f(x, y)$$

and equal to y_0 for $x = x_0$ has continuous derivatives with respect to x_0 and y_0 up to order $p \geqslant 1$. This assertion remains true for $p = 0$ if f is such as to guarantee uniqueness of the solution passing through (x_0, y_0).

Proof. Apply the theorem to equation (3.40). Also recall the theorem of Sec. 19 and the remark on p. 58.

Remark 1. Given a solution $\varphi(x, \mu_0)$ of (3.44) corresponding to some parameter value $\mu = \mu_0$, the derivative $\partial\varphi/\partial\mu$ for $\mu = \mu_0$ and for values of x and y satisfying the equation $y = \varphi(x, \mu_0)$ can be found directly (by quadratures) without finding the other solutions of (3.44). In fact, under these conditions, $\partial\varphi/\partial\mu$ is the solution of the linear equation (3.49) with known coefficients.

Remark 2. By using the method of Sec. 19, we can establish the continuous dependence of the solutions of (3.43) on the parameters $\mu_1, \ldots, \mu_n$ (or the differentiability of the solution with respect to these parameters) on any larger interval $[\alpha, \beta]$ containing $[a, b]$ such that the solution on $[\alpha, \beta]$ lies in $\bar{G}_{xy}$.

Problem 1. Given the equation $y' = \sin(xy)$ and the initial condition $x_0 = 0$, $y_0 = 0$, use (3.49) to find $\partial y/\partial x_0$ and $\partial y/\partial y_0$ (for arbitrary x).

Problem 2. Given the equation $y' = x^2 + y^2$ and the initial condition $x_0 = 0$, $y_0 = 0$, find the value of $\partial y/\partial y_0$ at $x = 1$ to within an error of 0.0001.

Hint. One way of finding $y(x)$ is to use the method of successive approximations.

Problem 3. Prove that if the right-hand side of $y' = f(x, y)$ is continuously differentiable, then

$$\frac{\partial y(x, x_0, y_0)}{\partial x_0} = -f(x_0, y_0) \exp\left\{ \int_{x_0}^{x} f_y[t, y(t)] \, dt \right\},$$

$$\frac{\partial y(x, x_0, y_0)}{\partial y_0} = \exp\left\{ \int_{x_0}^{x} f_y[t, y(t)] \, dt \right\}.$$

Problem 4. Let the functions M and N appearing in equation (2.21), Sec. 8 be continuously differentiable, and suppose $M^2(x_0, y_0) + N^2(x_0, y_0) \neq 0$, where (x_0, y_0) is a point of the domain G. Prove that (2.21) has a continuous integrating factor in some neighborhood of (x_0, y_0).

Problem 5. Suppose that

a) $f(x, y)$ is continuous and satisfies a Lipschitz condition in y for $a \leqslant x \leqslant b$, $-\infty < y < \infty$;

b) $f(x, y) > 0$ for $y < F(x)$ and $f(x, y) < 0$ for $y > F(x)$, where $F(x)$ is some function continuous on $[a, b]$.

Then if $\mu > 0$, the solution $y(x, \mu)$ of the equation

$$\mu \frac{dy}{dx} = f(x, y)$$

satisfying the initial condition $y(x_0) = y_0$, where $a < x_0 < b$ and $y_0 > F(x_0)$, depends continuously on (x, μ) and is defined on $[a, b]$. Prove that

$$\lim_{\mu \to 0+} y(x, \mu) = +\infty \qquad (a \leqslant x < x_0),$$

$$\lim_{\mu \to 0+} y(x, \mu) = F(x) \qquad (x_0 < x \leqslant b),$$

where the second formula holds uniformly in x on every interval $[c, b]$ $(x_0 < c < b)$.

Comment. Thus the way the solution of a first-order differential equation depends on changes of a small parameter multiplying y' is essentially different from the way it depends on changes of a small parameter appearing in the right-hand side of the equation. The above result is a special case of a much more general theorem due to Tikhonov.[33]

---

[33] A. N. Tikhonov, *On the dependence of solutions of differential equations on a small parameter* (in Russian), Mat. Sb., **22**, 193 (1948); *On systems of differential equations containing parameters* (in Russian), ibid., **27**, 147 (1950).

22. Singular Points

Consider the differential equation

$$\frac{dy}{dx} = f(x, y) \tag{3.50}$$

or

$$\frac{dx}{dy} = f_1(x, y) = \frac{1}{f(x, y)} \tag{3.50'}$$

(see Sec. 2) defined on a domain and possibly on some or all of the boundary of G. Then a point P in $\bar{G}$ is said to be an *ordinary point* of equation (3.50), (3.50') if there is a neighborhood $\mathcal{N}$ of P such that[34]

a) One and only one integral curve of (3.50), (3.50') passes through each point of $\mathcal{N}$;

b) At least one of the functions $f(x, y)$, $f_1(x, y)$ is continuous on $\mathcal{N}$.

Otherwise P is said to be a *singular point* of equation (3.50), (3.50').

A sufficient condition for P to be an ordinary point is that $f(x, y)$ be continuous in x and satisfy a Lipschitz condition in y on $\mathcal{N}$, or that $f_1(x, y)$ be continuous in y and satisfy a Lipschitz condition in x on $\mathcal{N}$.[35] However this condition is not necessary, as shown by the equation

$$y' = \begin{cases} y \ln |y| & \text{for } y \neq 0, \\ 0 & \text{for } y = 0. \end{cases}$$

In this case, every point of the plane is an ordinary point, although the Lipschitz condition breaks down on the x-axis.

The point P is singular in each of the following cases:

1. P belongs to the boundary of G. For example, the origin is a singular point of equations (1.4) and (1.7), since in both cases G is the whole plane except for the origin.

2. P is a "point of nonuniqueness" (i.e., two or more integral curves may pass through P in every neighborhood of P), or P is a limit point of points of nonuniqueness.

3. The given direction field has a discontinuity at P.[36]

---

[34] By a neighborhood $\mathcal{N}$ of a point P we shall always mean a "complete" neighborhood of P, e.g., a sufficiently small disk with P as its center, and not just the intersection of such a disk with the domain G. Neighborhoods of other shapes will be allowed (cf. pp. 76, 170).

[35] Instead of a Lipschitz condition, we might require that the corresponding partial derivative be bounded on $\mathcal{N}$.

[36] This case is seldom encountered in practice (however, see Sec. 23).

Remark. Naturally, combinations of these cases (e.g., a boundary point of nonuniqueness) are possible.

By an *isolated singular point* we mean a singular point which has a neighborhood containing no other singular points. These are most frequently encountered in studying equations of the form

$$\frac{dy}{dx} = \frac{M(x, y)}{N(x, y)} \qquad (3.51)$$

or

$$\frac{dx}{dy} = \frac{N(x, y)}{M(x, y)}, \qquad (3.51')$$

where M and N have continuous partial derivatives with respect to x and y of sufficiently high orders. Clearly, any point at which M and N do not vanish simultaneously is an ordinary point of equation (3.51), (3.51'). If on the other hand, (x_0, y_0) is a point where M and N both vanish, we assume for simplicity that $x_0 = y_0 = 0$, and then expand M and N in powers of x and y up to terms of order two, obtaining

$$\frac{dy}{dx} = \frac{M_x(0, 0)x + M_y(0, 0)y + O(x^2 + y^2)}{N_x(0, 0)x + N_y(0, 0)y + O(x^2 + y^2)} \qquad (3.52)$$

in a neighborhood of the origin.[37] This equation does not define dy/dx at the origin. Moreover, dy/dx does not approach any limit as $(x, y) \to (0, 0)$ if

$$\begin{vmatrix} M_x(0, 0) & M_y(0, 0) \\ N_x(0, 0) & N_y(0, 0) \end{vmatrix} \neq 0$$

(see below), and hence no definition of $M(0, 0)/N(0, 0)$ makes $M(x, y)/N(x, y)$ continuous at the origin, i.e., the origin is an (isolated) singular point of equation (3.51), (3.51').

Perron has shown[38] that the terms $O(x^2 + y^2)$ appearing in the numerator and denominator of (3.52) have no effect at all on the behavior of the integral curves near the isolated singular point (here the origin), provided that the real parts of both roots of the equation

$$\begin{vmatrix} \lambda - N_x(0, 0) & -N_y(0, 0) \\ -M_x(0, 0) & \lambda - M_y(0, 0) \end{vmatrix} = 0$$

are nonzero (cf. Sec. 56). Therefore to get an idea of how the integral curves

---

[37] By $O(x^2 + y^2)$ is meant a quantity whose ratio to $x^2 + y^2$ remains bounded [as $(x, y) \to (0, 0)$].

[38] O. Perron, *Über die Gestalt der Integralkurven einer Differentialgleichung ersten Ordnung in der Umgebung eines singulären Punktes*, Math. Z., **15**, 121 (1922); *Part 2*, ibid., **16**, 273 (1923). See also I. Bendixson, *Sur les courbes définies par des équations différentielles*, Acta Math., **24**, 1 (1901) and M. Frommer, *Über das Auftreten von Wirbeln und Strudeln* (*geschlossener und spiraliger Integralkurven*) *in der Umgebung rationaler Unbestimmtheitsstellen*, Math. Ann., **109**, 395 (1934).

behave near the origin, we need only study the integral curves of the equation

$$\frac{dy}{dx} = \frac{ax + by}{cx + dy}, \tag{3.53}$$

where

$$\begin{vmatrix} a & b \\ c & d \end{vmatrix} \neq 0.$$

It can be shown (see Sec. 48 below) that a nonsingular linear transformation

$$x = k_{11}\xi + k_{12}\eta,$$
$$y = k_{21}\xi + k_{22}\eta$$

with real k_{ij} reduces (3.53) to one of the following three forms:

$$\frac{d\eta}{d\xi} = k\frac{\eta}{\xi} \qquad (k \neq 0), \tag{3.54}$$

$$\frac{d\eta}{d\xi} = \frac{\xi + \eta}{\xi}, \tag{3.55}$$

$$\frac{d\eta}{d\xi} = \frac{\xi + k\eta}{k\xi - \eta}. \tag{3.56}$$

We now consider each of these cases in detail, noting first that although the x and y-axes intersect at right angles, the same is not true in general of the ξ and η-axes. However, to keep things simple, the ξ and η-axes will be represented by perpendicular lines in the figures.

Case 1. A complete integral of (3.54) is

$$a\eta + |b\xi|^k = 0,$$

and the corresponding behavior of the integral curves is shown schematically in Figures 12, 13 and 1. Figure 12 pertains to the case where $k > 1$. Here, except for the two halves of the η-axis, all the integral curves are tangent to

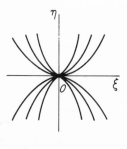

FIGURE 12

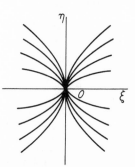

FIGURE 13

the ξ-axis at the origin. The ξ and η-axes are integral curves, except of course at the origin itself, where equation (3.54) defines no direction at all. The case $k = 1$ has already been treated in Chap. 1 (see Figure 1, p. 7). If $0 < k < 1$, all the integral curves except the two halves of the ξ-axis are tangent to the η-axis, as shown in Figure 13.

Examining the figures, we see that if $k > 0$, every integral curve approaches the origin O along a definite direction (i.e., has a definite tangent at O). In general, a point P is called a *node* if every integral curve with points sufficiently close to P approaches P along a definite direction. Thus, if $k > 0$, the point O is a node for the integral curves of equation (3.54).

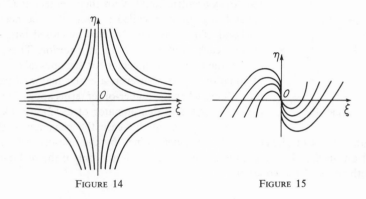

FIGURE 14   FIGURE 15

The integral curves

$$\eta \, |\xi|^{-k} = c$$

corresponding to the case $k < 0$ are shown in Figure 14. Now only four integral curves approach the point O arbitrarily closely, i.e., the two halves of the ξ-axis and the two halves of the η-axis. Every other integral curve comes to within a certain distance of O and then begins to recede from O. Singular points of this kind are called *saddle points* (the integral curves resemble the level lines on a map corresponding to a mountain pass).

Case 2. A complete integral of (3.55) is

$$b\eta = \xi(a + b \, |\ln \xi|),$$

and all the integral curves are tangent to the η-axis at the origin O (see Figure 15). This time the η-axis but not the ξ-axis is an integral curve, and the point O is again a *node*, as in Case I for $k > 0$.

Case 3. Equation (3.56) is most easily integrated by going over to polar coordinates. Thus, writing

$$\xi = \rho \cos \varphi, \qquad \eta = \rho \sin \varphi,$$

we find that (3.56) is equivalent

$$\frac{d\rho}{d\varphi} = k\rho,$$

and hence

$$\rho = Ce^{k\varphi}.$$

If $k > 0$, all the integral curves approach O winding around O infinitely often as $\varphi \to -\infty$, while if $k < 0$, the same thing happens as $\varphi \to +\infty$

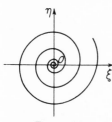

FIGURE 16

(see Figure 16). In either case, O is called a *focus*. If $k = 0$, the family of integral curves of (3.56) consists of concentric circles with their centers at O. In general, a point P is called a *center* if some neighborhood of P is completely filled by closed integral curves, each containing P in its interior. Thus, if $k = 0$, the point O is a center for the integral curves of equation (3.56). However, a center may well turn into a focus when the higher-order terms appearing in the numerator and denominator of (3.52) are taken into account, i.e., in this case the behavior of the integral curves of (3.51), (3.51′) is not determined by the first-order terms. Later on, in Sec. 48, we shall see that k vanishes if and only if the real parts of both roots of the equation

$$\begin{vmatrix} \lambda - c & -d \\ -a & \lambda - b \end{vmatrix} = 0$$

vanish. This cannot happen in any of the other cases considered above.

The classification of singular points given in this section is due to Poincaré.

Problem 1. Examine the behavior of the integral curves of the following equations near the appropriate singular points:

a) $y' = \dfrac{y}{x^2}$;   b) $y' = \dfrac{x}{y^2}$;   c) $y' = \dfrac{x^2}{y^2}$;

d) $y' = \dfrac{y^2}{x^2}$;   e) $y' = x^m y^n$ (for integral m and n).

Problem 2. Examine the behavior of the integral curves of the equation

$$y' = \left(\sin \frac{1}{x}\right)^{\pm 1} \left(\sin \frac{1}{y}\right)^{\pm 1}$$

(there are four distinct cases).

Problem 3. How do the integral curves of equation (3.53) behave if

$$\begin{vmatrix} a & b \\ c & d \end{vmatrix} = 0?$$

Problem 4. Give an example of a singular point P with the property that a unique integral curve passes through each point of some neighborhood of P.

Problem 5. For what values of a, b, c and d does equation (3.53) go into itself a) under all homogeneous affine transformations; b) under all rotations?

23. Singular Curves

DEFINITION. *If every point of a curve is a singular point [of equation (3.50), (3.50′)], the curve is said to be a singular curve. If a singular curve is also an integral curve, it is said to be a singular integral curve.*

Example 1. Every curve which is not an integral curve of the equation

$$y' = f(x, y),$$

where $f(x, y)$ is the function constructed by Lavrentev and alluded to in Sec. 9, is a singular curve (but obviously not a singular integral curve).

Example 2. The x-axis is a singular integral curve of the equation

$$y' = \begin{cases} y \ln^2 |y| & \text{for} \quad y \neq 0, \\ 0 & \text{for} \quad y = 0, \end{cases}$$

in fact an "integral curve of nonuniqueness." In practice, it is usually important to find such curves, since they help to form a picture of the behavior of the integral curves "in the large" (see below).

Example 3. Let G be the domain on which at least one of the functions $f(x, y), f_1(x, y)$ appearing in (3.50), (3.50′) is defined. Then a curve is a singular curve of (3.50), (3.50′) if it is part of the boundary of G. Such a curve may also be an integral curve, if $f(x, y)$ or $f_1(x, y)$ is defined on G. Then we have a "boundary integral curve."

Example 4. The x-axis is an integral curve of the equation

$$y' = \operatorname{sgn} y \equiv \begin{cases} 1 & \text{for} \quad y > 0, \\ 0 & \text{for} \quad y = 0, \\ -1 & \text{for} \quad y < 0, \end{cases}$$

since in this case the direction field is discontinuous at every point of the curve.

24. Behavior of Integral Curves in the Large. Limit Cycles

It is often important to form a qualitative picture of the behavior of the integral curves of a differential equation "in the large," i.e., of their behavior on the whole domain of definition of the direction field, without worrying about the detailed geometry of the problem. For example, Figures 12–16 depict the behavior of integral curves near an isolated singular point. Let G be a simply connected domain on which the right-hand side of equation (3.50), (3.50') is defined, and suppose every point of G is an ordinary point. Then the family of integral curves can be represented schematically by a family of parallel line segments, since in this case an integral curve passes through every point of G and no two integral curves intersect.

The structure of the integral curves can be much more complicated if equation (3.50), (3.50') has singular points or singular curves. One of the fundamental problems of the theory of differential equations is to find out (by the simplest possible means) how the family of integral curves of a given differential equation behaves on the whole domain G, i.e., to study the behavior of the integral curves "in the large."[39] A definitive solution of this problem is still not available even for equations of the form

$$\frac{dy}{dx} = \frac{M(x, y)}{N(x, y)},$$

where $M(x, y)$ and $N(x, y)$ are polynomials of degree higher than two.

Having made these general remarks, we now turn our attention to an important concept of the qualitative theory of differential equations, i.e., that of the limit cycle. Consider the differential equation

$$\frac{d\rho}{d\varphi} = \rho - 1, \tag{3.57}$$

where ρ and φ are polar coordinates in the xy-plane.[40] The complete integral

---

[39] The adjectives "global" and "nonlocal" are synonyms for the phrase "in the large." The branch of differential equations which deals with global behavior of integral curves is called the "qualitative" or "geometric" theory of differential equations. See e.g., V. V. Nemytski and V. V. Stepanov, *Qualitative Theory of Differential Equations*, Princeton University Press, Princeton, N.J. (1960), and S. Lefschetz, *Differential Equations: Geometric Theory*, second edition, Interscience Publishers, New York (1962).

[40] In rectangular coordinates, (3.57) takes the form

$$\frac{dy}{dx} = \frac{(x + y)\sqrt{x^2 + y^2} - y}{(x - y)\sqrt{x^2 + y^2} - x}.$$

of (3.57) is

$$\rho = 1 + Ce^{\varphi},$$

where C is an arbitrary constant and φ cannot take values larger than $-\ln |C|$ if $C < 0$ (otherwise ρ becomes negative). The corresponding family of integral curves (see Figure 17) consists of

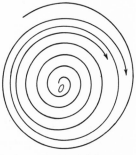

1) The circle $\rho = 1$ ($C = 0$);

2) Spirals emanating from the origin O and approaching the circle $\rho = 1$ from the inside as $\varphi \to -\infty$ ($C < 0$);

3) Unbounded spirals approaching the circle $\rho = 1$ from the outside as $\varphi \to -\infty$ ($C > 0$).

FIGURE 17

A closed integral curve is said to be a *limit cycle* (of a given differential equation) if it consists entirely of ordinary points and if some other integral curve approaches it asymptotically.[41] Thus the circle $\rho = 1$ is a limit cycle of equation (3.57). The search for limit cycles is a problem of great physical interest.

Remark. The fact that every point of the circle $\rho = 1$ is an ordinary point of equation (3.57) can be verified by transforming from polar coordinates to rectangular coordinates. Thus a small neighborhood of any point of the limit cycle is in no way different from a small neighborhood of any other nonsingular point.

Problem 1. Show that a necessary and sufficient condition for a direction field defined on a domain G to be representable by equations of the form (3.51), (3.51′), where M and N are continuous and do not vanish simultaneously, is that it be possible to equip every point of G with a unit vector which coincides with the direction of the field and depends continuously on position in the field. As before, the direction of the field at each point is specified by a line segment, but now (cf. footnote 2, p. 5) one of the directions of the segment must be chosen at each point and this direction must be a continuous function of position.

Problem 2. Give an example of a direction field on a (two-dimensional) annulus which cannot be represented on the whole annulus by an equation of the form (2.21), where M and N are continuous and do not vanish simultaneously, but which nevertheless varies continuously on the whole annulus. As usual, we

---

[41] Another definition of a limit cycle encountered in the literature is the following: A closed integral curve L is said to be a limit cycle if it consists entirely of ordinary points and if there are no other closed integral curves in some strip containing L.

represent the direction field at every point by a line segment, without distinguishing between the two directions of the segment, and in saying that the direction field varies continuously on the annulus we mean that this segment varies continuously. Can an analogous example be constructed for a simply connected domain in the plane?

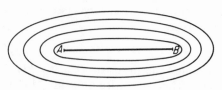

FIGURE 18

Problem 3. Suppose the integral curves of equation (3.51), (3.51') have the form shown in Figure 18. Prove that if M and N are continuous, then they must vanish everywhere on the segment AB.

Problem 4. Prove that the number of integral curves of equation (3.51), (3.51') which can enter an isolated singular point must either be even or infinite. Prove that if any integral curve approaches an isolated singular point along a spiral winding around the point an infinite number of times, then every other integral curve must approach the point along such a spiral. Can precisely two such spirals approach an isolated singular point?

Problem 5. Sketch the behavior of the integral curves of the equation

$$\frac{dy}{dx} = -\frac{x}{y} + x^2 + y^2 - 1$$

in the whole plane. Show that there is a focus at the origin and a limit cycle with equation $x^2 + y^2 = 1$.

Hint. Compare the slope of the integral curve of this equation with the slope of the integral curves of the equation

$$\frac{dy}{dx} = -\frac{x}{y}$$

at the same points.

Problem 6. Sketch the behavior of the integral curves of the equation

$$\frac{dy}{dx} = \frac{y - x^2}{y - x} + (y - x)^2 + (x - 1)^2 + \frac{2}{3}(x - 1)^3 - \frac{1}{3}$$

in the whole plane. Show that there are two singular points, a saddle point at $(0, 0)$ and a focus at $(1, 1)$.

Hint. Compare the slope of the integral curves of this equation with the slope of the integral curves of the equation

$$\frac{dy}{dx} = \frac{y - x^2}{y - x}$$

(which is easily solved by quadratures) at the same points.

Problem 7. Prove that if a closed integral curve L consisting entirely of ordinary points can be included in a strip containing no other closed integral curves, then L is a limit cycle.

Problem 8. Prove that if a closed integral curve L containing no ordinary points is approached asymptotically from the inside and from the outside by two integral curves, then L can be included in a strip entirely filled with integral curves approaching L asymptotically.

Problem 9. Construct an example of a closed integral curve L without singular points which is not a limit cycle, where no neighborhood of L is entirely filled with closed integral curves.

Problem 10. Prove that if a direction field has no singular points inside or on a closed curve L with a continuously varying tangent, then the direction of the field coincides at least twice at points of L with the direction of the tangent to L and at least twice with the direction of the normal to L.

Comment. In particular, this implies a theorem of Bendixson which states that there must be at least one singular point of the field inside a closed integral curve.

Problem 11. Prove that every integral curve of the equation

$$\frac{dy}{dx} = x^2 - y^2$$

has at least one inflection point and intersects the line $y = x$ at least once.

Hint. Use Prob. 5, Sec. 2.

25. Equations Not Solved for y'

We now consider first-order differential equations of the general form

$$F(x, y, y') = 0,$$

rather than of the special form

$$y' = f(x, y).$$

Example. The equation

$$y'^2 - 1 = 0 \qquad (3.58)$$

is equivalent to the pair of equations

$$y' = +1, \qquad (3.59)$$

$$y' = -1. \qquad (3.59')$$

Every direction in the field corresponding to (3.59) makes an angle of 45° with respect to the x-axis, while every direction in the field corresponding to (3.59') makes an angle of 135° with respect to the x-axis. Therefore, given

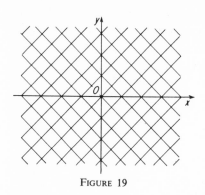

FIGURE 19

any point (x_0, y_0) of the xy-plane, there is one and only one integral curve of (3.59) passing through (x_0, y_0) [a line of slope $+1$] and one and only one integral curve of (3.59′) passing through (3.59′) [a line of slope -1]. The direction field of (3.58) is obtained by superposition of the direction fields of (3.59) and (3.59′), and as a result, there are two and only two integral curves of (3.58) passing through each point (x_0, y_0) of the xy-plane, as shown in Figure 19.[42]

This example is a special case of the following general

THEOREM.   *Given a differential equation*

$$F(x, y, y') = 0, \tag{3.60}$$

suppose $F(x, y, y')$ is such that

1) *$F(x, y, y')$ is defined and continuous on a closed bounded domain $\bar{G}$ in (x, y, y') space;*

2) *At some point (x_0, y_0) of the xy-plane, equation (3.60) has a finite number of distinct roots $b_1, \ldots, b_m$ when solved for y';*

3) *Each of the points (x_0, y_0, b_i), $i = 1, \ldots, n$ belongs to G and has a neighborhood R_i in which $F(x, y, y')$ has continuous partial derivatives $F_y(x, y, y')$ and $F_{y'}(x, y, y')$, where $|F_{y'}(x, y, y')| \geqslant c > 0$.*

Then there is a neighborhood $\mathcal{N}$ of (x_0, y_0) such that precisely m solutions of (3.60) pass through each point of $\mathcal{N}$.

Proof.   Under these assumptions, it follows from the implicit function theorem that each of the points (x_0, y_0, b_i) has a complete neighborhood R_i in (x, y, y') space (cf. footnote 34, p. 66) in which (3.60) has one and only one solution of the form

$$y' = f_i(x, y), \qquad i = 1, \ldots, m, \tag{3.61}$$

---

[42] By the intermediate value theorem for derivatives (see e.g., T. M. Apostol, *op. cit.*, p. 94), if $\Phi(x)$ has a derivative $\varphi(x)$ everywhere in an interval (a, b) and if $\varphi(x)$ takes the values y_1 and y_2 at points x_1 and x_2 in (a, b), then $\varphi(x)$ takes every value between y_1 and y_2 at some point of (x_1, x_2). Hence there is no function $y(x)$ whose derivative $y'(x)$ exists for all x and takes only the values ± 1, such that $y'(x) = +1$ for some values of x and $y'(x) = -1$ for other values of x. The fact that there are no smooth integral curves other than those mentioned is also obvious from Figure 19.

where $f_i(x, y)$ is continuous in x and has a derivative with respect to y
equal to

$$- \frac{F_y(x, y, f_i)}{F_{f_i}(x, y, f_i)}, \qquad i = 1, \dots, m. \tag{3.62}$$

Since $|F_{f_i}(x, y, f_i)| \geqslant c > 0$, the quantities (3.62) are bounded. The
neighborhoods $R_1, \dots, R_m$ can be represented as cylinders with
generators parallel to the y'-axis, where the
projection of the base of each cylinder onto
the xy-plane is the same neighborhood $\mathcal{N}$
of the point (x_0, y_0), as shown in Figure
20 for the case $m = 2$. This neighborhood
$\mathcal{N}$ can be chosen so small that there is no
point (x, y, y') of the surface (3.60) either
above or below $\mathcal{N}$ which does not belong
to one of the surfaces (3.61). In fact,
if such points existed for arbitrarily small
$\mathcal{N}$, then, since $F(x, y, y')$ is continuous
and $\bar{G}$ is closed and bounded, they would
have to lie outside the cylinders $R_1, \dots, R_m$
on the line $x = x_0$, $y = y_0$. But this is impossible, since then (3.60) would have more
than m roots when solved for y' at the point
(x_0, y_0).

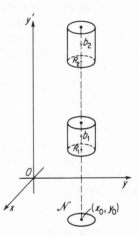

FIGURE 20

Thus we find that the point (x_0, y_0) has a neighborhood $\mathcal{N}$ (in the xy-
plane) where equation (3.60) has precisely m solutions (3.61). Each
function $f_i(x, y)$, $i = 1, \dots, m$ is continuous in x and has a bounded
derivative with respect to y. Therefore each of the equations (3.61) has
one and only one integral curve passing through any given point of $\mathcal{N}$.
Since the y' are all different in $\mathcal{N}$, these integral curves are all different
and no two make contact without intersecting. Therefore, as asserted,
precisely m integral curves of equation (3.60) pass through each point of
$\mathcal{N}$ (cf. footnote 42, p. 76).

Obviously, none of the directions of the field defined by (3.60) is parallel
to the y-axis, and hence none of the integral curves of (3.60) can have
tangents parallel to the y-axis. However, just as in the case of equations solved
for y', we would like to include the possibility of directions parallel to the
y-axis. Therefore, besides the equation

$$F\left(x, y, \frac{dy}{dx}\right) = 0, \tag{3.63}$$

we shall sometimes consider the equation

$$F_1\left(x, y, \frac{dx}{dy}\right) = 0, \tag{3.63'}$$

where the function F_1 is chosen in such a way that the equations (3.63) and (3.63′) are consistent. It is sometimes more convenient to combine (3.63) and (3.63′) into a single equation written in terms of differentials (see Example 1 below). Thus, just as in Sec. 2, in addition to solutions of equations of the form (3.60), we shall consider integral curves of the pair of equations (3.63) and (3.63′), or more concisely, of the equation (3.63), (3.63′).

Now suppose the equation (3.63), (3.63′) is defined on a domain $G_{xyy'}$ [or $G_{xyx'}$] in (x, y, y') space [or (x, y, x') space] and possibly on some or all of the boundary of the domain. Let G_{xy} be the domain in the xy-plane where equation (3.63), (3.63′) defines a direction field. Then a point P in $\bar{G}_{xy}$ is said to be an *ordinary point* of equation (3.63), (3.63′) if there is a neighborhood $\mathcal{N}$ of P in the xy-plane such that the same finite number of integral curves, equal to the number of directions specified by (3.63), (3.63′) at P, passes through each point of $\mathcal{N}$. These curves are obtained by superposition of families of integral curves of equations of the form (3.50), (3.50′), where at least one of the functions $f(x, y)$, $f_1(x, y)$ is continuous on $\mathcal{N}$. Otherwise, P is said to be a *singular point* of equation (3.63), (3.63′).

According to the theorem just proved, the following are sufficient conditions for P to be an ordinary point of (3.63), (3.63′):

a) Given any set E in the xy-plane, let $(E)_{xyy'}$ [or $(E)_{xyx'}$] be the set of all points (x, y, y') [or (x, y, x')] such that (x, y) belongs to E and $F(x, y, y')$ [or $F_1(x, y, x')$] is defined. Then the point $P = (x_0, y_0)$ has a closed neighborhood $\mathcal{N}$ such that $\bar{G}^*_{xyy'} = (\bar{\mathcal{N}})_{xyy'}$ [or $\bar{G}^*_{xyx'} = (\bar{\mathcal{N}})_{xyx'}$] is closed and bounded, and $F(x, y, y')$ [or $F_1(x, y, x')$] is continuous on $\bar{G}^*_{xyy'}$ [or $\bar{G}^*_{xyx'}$].

b) The number of integral curves specified by equation (3.63), (3.63′) at the point P is finite.

c) For each of the directions specified by (3.63) [or (3.63′)], the function $F(x, y, y')$ [or $F_1(x, y, x')$] satisfies condition 3 of the theorem (on p. 76) at the point P.[43]

Thus, if F and F_1 are sufficiently smooth, the singular points of equation (3.63), (3.63′) must either be boundary points, or else satisfy one of the systems

$$F(x, y, y') = 0, \qquad F_{y'}(x, y, y') = 0, \qquad (3.64)$$

$$F_1(x, y, x') = 0, \qquad F_{1x'}(x, y, x') = 0, \qquad (3.64')$$

where it is assumed that conditions a and b are met.

Finally, we define *singular curves* and *singular integral curves* just as in Sec. 23, except that we now start from the concept of a singular point of the general equation (3.63), (3.63′), rather than of a singular point of equation

---

[43] In the case of $F_1(x, y, x')$, the roles of x and y, and of y' and x', must be reversed in the statement of the theorem.

(3.50), (3.50′) which is a special case of (3.63), (3.63′). The examples of singular points, singular curves and singular integral curves considered in Secs. 22 and 23 still apply here. We now present two further examples, based on equations of the form (3.63), (3.63′).

Example 1. Consider the equations

$$y'^2(1 - x^2) - x^2 = 0, \tag{3.65}$$

$$(1 - x^2) - x^2 x'^2 = 0, \tag{3.65'}$$

which can be written in the more symmetrical form

$$(1 - x^2)(dy)^2 - x^2(dx)^2 = 0. \tag{3.66}$$

Equation (3.66) defines a direction field only in the strip $|x| \leqslant 1$. The left-hand side of (3.65) is continuous and has continuous derivatives with respect to y and y' everywhere in this strip. Its derivative with respect to y' is $2y'(1 - x^2)$ and vanishes only if $x = \pm 1$ or $y' = 0$. Because of (3.65), $y' = 0$ implies $x = 0$. The derivative with respect to x' of the left-hand side of (3.65′) vanishes on the same lines (and nowhere else), and hence (3.66) can have no singular curves other than the three lines

$$x = -1, \quad x = 0, \quad x = +1.$$

The lines $x = \pm 1$ are singular curves, since they form the boundary of the domain where (3.66) defines a direction field. Moreover, it is clear from (3.65′) that the lines $x = \pm 1$ are integral curves.

Next we show that line $x = 0$ is also a singular curve (but not an integral curve). First we observe that (3.66) implies

$$\frac{dy}{dx} = \pm \frac{x}{\sqrt{1 - x^2}},$$

and hence all circles of radius 1 with centers on the y-axis are integral curves of (3.66) [note that all these circles are tangent to the lines $x = \pm 1$]. It is then easy to see that the line $x = 0$ is a singular curve. In fact, on this line equation (3.65) has only *one* solution $y' = 0$, while equation (3.65′) cannot be solved for x' at all. But there is no point B on the y-axis with a neighborhood $\mathcal{N}$ such that one and only one integral curve passes through each point of $\mathcal{N}$, since there are *four*

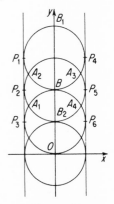

FIGURE 21

integral curves passing through B itself, i.e., the curves $A_1 B A_4$, $A_2 B A_3$, $A_2 B A_4$ and $A_1 B A_3$ shown in Figure 21. In other words, the y-axis is a singular curve, but obviously not an integral curve. On the other hand, if P is any point of the strip $-1 < x < 1$ which does not lie on the y-axis, then two and only two integral curves pass through each point of some neighborhood of P.

Finally we note that besides the integral curves already mentioned, equation (3.66) also has integral curves of the form

$$P_1B_1P_4P_5B_2P_2P_1, \qquad P_1B_1P_4BP_3OP_6BP_1,$$

and so on. Thus infinitely many curves pass through any point of the lines $x = \pm 1$, and hence these lines might be called "boundary integral curves of nonuniqueness."

Example 2. Consider the equation

$$F(x, y, y') \equiv y - xy' - f(y') = 0, \tag{3.67}$$

known as *Clairaut's equation.* We shall assume that $f(y')$ is continuous together with its first and second derivatives on a closed interval $a \leqslant y' \leqslant b$, and that $f''(y')$ has constant sign (say negative) on this interval. Then for every x and y equation (3.67) has no more than two roots when solved for y' (why?). Therefore it is easy to see that (x_0, y_0) is an ordinary point of (3.67) if

1) The equation

$$y_0 - x_0 y' - f(y') = 0 \tag{3.68}$$

is satisfied for at least one value of y' in the open interval (a, b) but not for $y' = a$ or $y' = b$;

2) $F_{y'}(x_0, y_0, y') = -x_0 - f'(y') \neq 0$, where y' is one of the roots of (3.68).

The points which do not satisfy condition 2 form a curve, which, in terms of the parameter $p = y'$, can be written as

$$x = -f'(p), \qquad y = xp + f(p) \tag{3.69}$$

or

$$x = -f'(p), \qquad y = -f'(p)p + f(p). \tag{3.70}$$

These equations define y as a function of x. To see this, we need only solve $x = -f'(p)$ for p [which is possible, since $f''(p)$ has constant sign], and then substitute the result into the equation for y. Clearly, the solution of (3.70) is an integral curve. In fact, it follows from (3.70) that

$$dx = -f''(p),$$
$$dy = [-f''(p)p - f'(p) + f'(p)]\, dp = -pf''(p)\, dp,$$

and hence

$$\frac{dy}{dx} = p, \tag{3.71}$$

i.e., the solution of (3.70), or equivalently of (3.69), has the slope given by
(3.67). Because of (3.71), if p increases, so does dy/dx. Moreover, (3.70)
shows that

$$\frac{dx}{dp} > 0$$

if

$$f''(p) < 0.$$

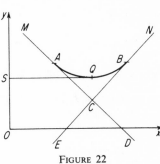

FIGURE 22

Therefore the curve (3.70) is concave upward, as shown in Figure 22, where the
curve (3.70) is indicated by AQB.

It is easily verified that the line

$$y = cx + f(c) \qquad (3.72)$$

is an integral curve for any constant c $(a \leqslant c \leqslant b)$. Clearly, this line is
tangent to the curve (3.70) at the point

$$x = -f'(c), \qquad y = -f'(c)c + f(c).$$

Moreover, every tangent to the curve (3.70) is of the form (3.72) [why?].
Therefore equation (3.68) has as many roots (when solved for y') as there are
tangents to the arc AQB passing through the point (x_0, y_0).

We now draw the tangents $MACD$ and $NBCE$ to the arc AQB at its end
points A and B. Together with AQB, these lines divide the whole xy-plane
into five domains

$$MACE, \quad NBCD, \quad AQBCA, \quad ECD, \quad MAQBN.$$

There is one and only one integral curve tangent to the arc AQB passing
through each point P inside the angles $MACE$ and $NBCD$, and hence
equation (3.67) has one and only one root y' (distinct from a and b) at each
such point P, which is therefore an ordinary point. In other words, every
point P in these domains has a neighborhood $\mathcal{N}$ such that the number of
integral curves of (3.67) passing through each point of $\mathcal{N}$ is the same as the
number of roots of (3.67) at P, i.e., exactly *one*. This unique integral curve is
a segment of the tangent line to AQB drawn through P. In the same way, we
find that every point inside the domain $AQBCA$ is also an ordinary point
of (3.67). In fact, every point P in this domain has a neighborhood $\mathcal{N}$ such
that two and only two integral curves pass through each point of $\mathcal{N}$, where
the integral curves are again segments of tangents to AQB drawn through P.
On the other hand, every interior point of the domains $MAQBN$ and ECD
has a neighborhood containing no integral curves of (3.67), and hence the
points of these domains are classified neither as ordinary points nor as
singular points.

Of all the points in the xy-plane for which equation (3.67) has roots when solved for y', only the points of the lines $MACD$ and $NBCE$ fail to satisfy condition 1 (on p. 80), and only the points of the curve AQB fail to satisfy condition 2. It is easy to see that these curves are singular curves of equation (3.67).

Finally we note that besides the integral curves mentioned above, equation (3.67) has further integral curves like the curve $SQBN$, consisting of parts of tangents to AQB and part of the arc AQB itself.

Problem 1. Determine the form of the integral curves of the following equations:

$$\text{a) } \sin y' = 0; \qquad \text{b) } \sin y' = x.$$

Problem 2. Consider the equation $F(x, y') = 0$. Prove that if the curve $F(x, z) = 0$ has a vertical tangent at $x = x_0$, $z = z_0$, which does not intersect the curve, then

1) There is an integral curve of $F(x, y) = 0$ passing through each point (x_0, y_0);

2) The integral curve has a cusp at (x_0, y).

Make a similar study of the points where the curve $F(x, z) = 0$ has a) maxima; b) cusps; c) self-intersections; d) vertical asymptotes. Analyze the following equations

$$\text{a) } (x^2 + y'^2)^2 = a^2(x^2 - y'^2); \qquad \text{b) } x(1 + y'^6) = y'^4.$$

Problem 3. Solve the Clairaut equation

$$y - xy' - y = 0.$$

Why does the family of integral curves have a different appearance from that of Figure 22?

Problem 4. Investigate the family of solutions of equation (3.67) if $f''(y')$, $a \leqslant y' \leqslant b$ vanishes a finite number of times.

Problem 5. One method of solving the system of equations

$$y = f(t), \qquad \frac{dy}{dx} = \varphi(t), \tag{3.73}$$

where t is a parameter, is to differentiate the first equation, which leads to

$$x = \int \frac{f'(t)}{\varphi(t)} \, dt + C. \tag{3.74}$$

Then (3.74) together with $y = f(t)$ gives a parametric representation of the family of solutions of (3.73). Give sufficient conditions for the validity of this method.

26. Envelopes

Suppose the differential equation $F(x, y, y') = 0$ has a family of integral curves

$$\Phi(x, y, C) = 0 \qquad (3.75)$$

covering a closed bounded domain $\bar{G}$ in the xy-plane in the sense that at least one curve of the family (3.75) passes through each point of $\bar{G}$, but only finitely many curves pass through any point of $\bar{G}$. Consider the problem of finding a curve L lying in $\bar{G}$ such that

1) L is tangent to some curve of the family (3.75) at each of its points;

2) Infinitely many curves of the family (3.75) are tangent to each arc of L.[44]

Such a curve L is called an *envelope* of the family (3.75). Obviously, the envelope L of a family of integral curves is itself an integral curve since at each of its points, L is tangent to an integral curve and hence has one of the field directions. We shall assume that *the function* $\Phi(x, y, C)$ *has continuous derivatives in all its arguments.* Further assumptions, indicated in italics, will be made later.

Suppose an envelope L exists. Then L is tangent at each of its points (x, y) to a curve L_C, where the index indicates the value of the parameter C for which the curve L_C is obtained from the general equation (3.75). Therefore the coordinates of the points of L satisfy the equation

$$\Phi[x, y, C(x, y)] = 0, \qquad (3.76)$$

where now C is no longer a constant, but varies along the curve L, taking the value of C corresponding to the variable L_C. We shall henceforth confine ourselves to arcs of L on which y *is a differentiable function of* x.[45] Then $C(x, y)$ in equation (3.76) can be regarded as a function of x only, so that (3.76) becomes

$$\Phi[x, y, C(x)] = 0. \qquad (3.77)$$

The function $C(x)$ is regarded as known, and moreover it is assumed that $C(x)$ *is differentiable and has no intervals of constancy* (i.e., no intervals on which it reduces to a constant).

---

[44] Curves of the family (3.75) are considered to be distinct if they correspond to distinct values of C. Regarded as geometric objects, two curves of the family (3.75) may coincide on some arc. However, because of our assumptions, it is impossible for all the curves of (3.75) tangent to an arc of an envelope of (3.75) to coincide on the arc.

[45] Arcs on which x is a differentiable function of y can be considered in just the same way.

We now find the value of y' for the function $y(x)$ satisfying (3.76). This can be done in two ways. First we can differentiate (3.77) with respect to x, obtaining

$$\frac{\partial \Phi}{\partial x} + \frac{\partial \Phi}{\partial y} y' + \frac{\partial \Phi}{\partial C} C' = 0. \tag{3.78}$$

On the other hand, the value of y' corresponding to the curve L_C passing through the same point (x, y) is found by differentiating (3.75) with respect to x:

$$\frac{\partial \Phi}{\partial x} + \frac{\partial \Phi}{\partial y} y' = 0. \tag{3.79}$$

Therefore in order for the values of y' given by (3.78) and (3.79) to coincide (y' can be determined from these equations *provided that* $\partial \Phi/\partial y \neq 0$), i.e., in order for the curves (3.75) and (3.76) to have a common tangent at the point (x, y), it is necessary that

$$\frac{\partial \Phi}{\partial C} C' = 0. \tag{3.80}$$

It follows from (3.80) that at least one of the factors $\partial \Phi/\partial C$ and C' vanishes. But C is assumed to have no intervals of constancy, and hence C' cannot vanish on any interval. Therefore the envelope satisfies the equation

$$\frac{\partial \Phi}{\partial C} = 0, \tag{3.81}$$

together with (3.77).

It is easily verified that conversely, if the equations (3.77) and (3.81) [where $\Phi(x, y, C)$ has the properties assumed above] determine a pair of differentiable functions $y(x)$ and $C(x)$, such that $C(x)$ has no intervals of constancy, then $y = y(x)$ is the equation of an envelope of the family (3.75).

Remark 1. Since x and y play identical roles in the statement of the problem, their roles can be reversed in its solution.

Remark 2. An envelope L of a family (3.75) of integral curves of a differential equation $F(x, y, y') = 0$ is always a singular integral curve of the equation. In fact, L is an integral curve, as already noted, and moreover all the points of L are singular, since infinitely many integral curves pass through any point P of L in an arbitrarily small neighborhood of P (why?)

Example 1. The family of curves

$$\Phi(x, y, C) \equiv y - (x + C)^3 = 0, \tag{3.82}$$

defined on the whole xy-plane, consists of the cubical parabolas obtained by subjecting one such parabola, say $y = x^3$, to shifts parallel to the x-axis.

Setting $\partial\Phi/\partial C$ equal to zero, we find that $-3(x + C)^2 = 0$ and hence $C = -x$. Substituting this value of C into (3.82), we obtain the line $y = 0$, which is obviously an envelope of the family (3.82), as shown in Figure 23. Note that if we had written the equation of the family in the form

$$\Phi(x, y, C) \equiv y^{1/3} - (x + C) = 0,$$

then $\partial\Phi/\partial C$ would equal -1 and our method would give no envelope, although one actually exists. This is due to the fact that now $\partial\Phi/\partial y$ does not exist for $y = 0$.

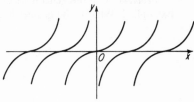

FIGURE 23

Example 2. Consider the family of curves

$$\Phi(x, y, C) \equiv y^5 - (x + C)^3 = 0, \tag{3.83}$$

defined on the whole xy-plane. Setting $\partial\Phi/\partial C$ equal to zero, we find that $-3(x + C)^2 = 0$ and hence $C = -x$. Substituting this value of C into (3.83), we obtain $y = 0$. But the x-axis is obviously not an envelope of the family (see Figure 24). This discrepancy is due to the fact that

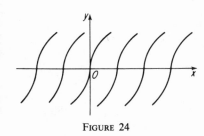

FIGURE 24

$$\frac{\partial\Phi}{\partial y} = 5y^4 = 0$$

for $y = 0$.

Example 3. The family of circles

$$\Phi(x, y, C) \equiv x^2 + (y + C)^2 - 1 = 0 \tag{3.84}$$

covers the strip between the lines $x = \pm 1$. Setting $\partial\Phi/\partial C$ equal to zero, we find that $2(y + C) = 0$ and hence $C = -y$. Substituting this value of C into (3.84), we obtain $x = \pm 1$. Each of these lines is an envelope of the family (3.84) [recall Figure 21].

Example 4. The equation

$$y - C^3 x^2 + 2C^2 x - C = 0, \tag{3.85}$$

which can be written in the form

$$y - C^3 \left(x - \frac{1}{C} \right)^2 = 0$$

if $C \neq 0$, defines a family of parabolas with axes parallel to the y-axis and vertices on the x-axis. The x-axis is obviously an envelope of this family. Moreover, the x-axis is itself a curve of the family, as can be seen by setting $C = 0$ in (3.85).

Problem. Use the general method of this section to find the envelopes in Example 4. What is the geometric meaning of the resulting hyperbola?

Part 2

SYSTEMS OF
ORDINARY DIFFERENTIAL EQUATIONS

4

GENERAL THEORY OF SYSTEMS

---

27. Reduction of an Arbitrary System to a System of First-Order Equations

Consider the system

$$\Phi_i\left(x, y_1, \frac{dy_1}{dx}, \ldots, \frac{d^{m_1} y_1}{dx^{m_1}}, \ldots, y_n, \ldots, \frac{d^{m_n} y_n}{dx^{m_n}}\right) = 0, \qquad (4.1)$$
$$i = 1, \ldots, n,$$

where each equation contains the independent variable x, n unknown functions $y_1, \ldots, y_n$ of x, and derivatives of each y_i with respect to x up to some order m_i. To reduce the system (4.1) to a first-order system, we write $y_i = y_i^{(0)}$ and

$$\frac{dy_i^{(k)}}{dx} = y_i^{(k+1)}, \qquad k = 0, 1, \ldots, m_i - 2. \qquad (4.2)$$

Then (4.1) takes the form

$$\Phi_i\left(x, y_1^{(0)}, y_1^{(1)}, \ldots, y_1^{(m_1-1)}, \frac{dy_1^{(m_1-1)}}{dx}, \ldots, y_n^{(0)}, y_n^{(1)}, \ldots, y_n^{(m_n-1)}, \frac{dy_n^{(m_n-1)}}{dx}\right) = 0,$$
$$i = 1, \ldots, n. \qquad (4.3)$$

Thus, given n functions $y_i(x)$, $i = 1, \ldots, n$ satisfying the system (4.1), we obtain a set of functions $y_i^{(k)}(x)$ which satisfy the system of first-order differential equations consisting of (4.2) and (4.3). Conversely, given a set of functions $y_i^{(k)}(x)$ satisfying (4.2) and (4.3), it is easy to see that the functions $y_i^{(0)}(x)$, $i = 1, \ldots, n$ satisfy the system (4.1). In fact, consecutively setting

$k = 0, 1, \ldots, m_i - 2$ in (4.2), we find that

$$y_i^{(k+1)} = \frac{d^{k+1} y_i^{(0)}}{dx^{k+1}},$$

and then (4.3) implies (4.1).

Henceforth we shall be concerned mainly with systems of first-order differential equations which are *solved with respect to the derivatives*.

> **Problem.** Prove that a system of first-order differential equations of the form
>
> $$f_i(x, y_1, \ldots, y_n, y_1', \ldots, y_n') = 0, \qquad i = 1, \ldots, n$$
>
> can "in general" be reduced by consecutive differentiation and elimination of superfluous variables to a single differential equation of order k involving one unknown function, whose solution allows the other unknown functions to be determined without further integrations. The phrase "in general" means that we assume the possibility of solving all relevant systems of finite (nondifferential) equations in which the number of equations equals the number of unknowns. It is assumed that the functions f_i have continuous derivatives of all required orders.

28. Geometric Interpretation. Definitions

Consider the system of differential equations

$$\frac{dy_i}{dx} = f_i(x, y_1, \ldots, y_n), \qquad i = 1, \ldots, n, \tag{4.4}$$

where the functions $f_i(x, y_1, \ldots, y_n)$ are defined on some domain G in $(x, y_1, \ldots, y_n)$ space.[1] A set of functions

$$y_1(x), \ldots, y_n(x) \tag{4.5}$$

satisfying the system (4.4) will be called a *solution* of the system. The equations

$$y_i = y_i(x), \qquad i = 1, \ldots, n \tag{4.6}$$

define a curve in $(x, y_1, \ldots, y_n)$ space called an *integral curve* (in fact, a "graph") of the system (4.4). Instead of saying that $y_i(x_0) = y_i^0, i = 1, \ldots, n$, we shall often say that the curve (4.6) or the solution (4.5) "passes through" the point $(x_0, y_1^0, \ldots, y_n^0)$. The set of functions

$$y_i = y_i(x, C_1, \ldots, C_m), \qquad i = 1, \ldots, n \tag{4.7}$$

involving parameters $C_1, \ldots, C_m$ is called the *general solution* of the system (4.4) in a domain G if every solution of (4.4) lying in G is given by (4.7) for a suitable choice of $C_1, \ldots, C_m$. The most common case is where $m = n$.

---

[1] I.e., the space of points $(x, y_1, \ldots, y_n)$.

The fact that the functions $y_i(x)$ satisfy the system (4.4) means geometrically that the tangent line l to the integral curve (4.6) at the point $(x, y_1(x), \ldots, y_n(x))$ is given by

$$\frac{Y_i - y_i(x)}{X - x} = f_i[x, y_1(x), \ldots, y_n(x)], \qquad i = 1, \ldots, n,$$

where X and Y_i denote running coordinates along l. Therefore the problem of solving the system (4.4) has the following geometric interpretation, which is the n-dimensional analogue of that given in Sec. 2: Suppose that through every point $(x, y_1, \ldots, y_n)$ of a domain G we draw a short line segment[2] whose direction cosines are proportional to the quantities

$$1, f_1(x, y_1, \ldots, y_n), \ldots, f_n(x, y_1, \ldots, y_n),$$

thereby specifying a set of directions in G, called the *direction field* of (4.4). Then finding an integral curve of the system (4.4) means finding a curve whose tangents have directions belonging to the given direction field.

If the direction field is specified in this way, none of the integral curves can lie in a plane parallel to the plane $x = 0$. The artificial character of this restriction is revealed by the following considerations: Starting from a *given* direction field in $(x, y_1, \ldots, y_n)$ space, we look for smooth curves[3] (called *integral curves* of the field) whose tangents have directions belonging to the field. In general, the equations of these curves cannot be written in the form (4.6), solved with respect to the unknowns $y_1, \ldots, y_n$, since planes parallel to the plane $x = 0$ may intersect the curves several times or even contain entire arcs of the curves. However, differential equations of the curves (or equivalently, of the corresponding direction field) can be written down by regarding x and the y_i as (continuously differentiable) functions of a suitable parameter t, e.g., the arc length along the integral curve or the time required to traverse the integral curve from some fixed point on the curve to a variable point $(x, y_1, \ldots, y_n)$. Then we have

$$\frac{dy_i}{dt} = f_i^*(x, y_1, \ldots, y_n), \qquad i = 1, \ldots, n,$$

$$\frac{dx}{dt} = f^*(x, y_1, \ldots, y_n), \tag{4.8}$$

where the functions f_i^* and f^* cannot all vanish simultaneously. The direction field represented by (4.8) can also be written in the form

$$\frac{dy_1}{f_1^*(x, y_1, \ldots, y_n)} = \cdots = \frac{dy_n}{f_n^*(x, y_1, \ldots, y_n)} = \frac{dx}{f^*(x, y_1, \ldots, y_n)}. \tag{4.9}$$

---

[2] Again we make no distinction between the two directions of the segment.

[3] Cf. Remark 1, p. 6.

On separate arcs of the curves, it may be possible to choose one of the coordinates $x, y_1, \ldots, y_n$ as the parameter. For example if $f_k^*(x, y_1, \ldots, y_n)$ is nonzero in some domain G, we can solve (4.9) for dx/dy_k and dy_i/dy_k $(i = 1, \ldots, k-1, k+1, \ldots, n)$, i.e., we can use y_k for the parameter. Similarly, if $f^*(x, y_1, \ldots, y_n)$ is nonzero in G, we can solve (4.9) for the dy_i/dx, and then x plays the role of the parameter.

The system (4.8) is of the form (4.4), except that the number of unknowns is one larger. Hence there is no loss of generality in confining ourselves to systems of the form (4.4). Such systems obey theorems analogous to those proved in Chapter 3, and in most cases the proofs are so similar that there is no need to repeat them in detail. Thus we shall merely state results, except in the case of Osgood's theorem and the principle of contraction mappings.

Remark. A system of equations

$$\Phi_i(x, y_1, \ldots, y_n) = 0, \qquad i = 1, \ldots, n$$

determining an integral curve of the system (4.4) is called an *integral* of (4.4). Moreover, a system of equations

$$\Phi_i(x, y_1, \ldots, y_n, C_1, \ldots, C_m) = 0, \qquad i = 1, \ldots, n \qquad (4.10)$$

involving parameters $C_1, \ldots, C_m$ is called a *complete integral* of the system (4.4) in a domain G if every integral curve of (4.9) lying in G is given by (4.10) for a suitable choice of $C_1, \ldots, C_m$.

Problem 1. Construct a system of two first-order differential equations of the form (4.4), with unknown functions y and z, whose integral curves are all right-handed helices with a given pitch h and the x-axis as common axis. How can this problem be generalized to a larger number of dimensions?

Problem 2. Give necessary and sufficient conditions for a direction field in a domain G to have a representation of the form (4.9), where the denominators are all continuous and do not vanish simultaneously. In what domains can any continuous direction field be represented in this form?

29. Basic Theorems

THEOREM 1 (*Peano's existence theorem*). *If the functions $f_i(x, y_1, \ldots, y_n)$ are continuous on a domain G in $(x, y_1, \ldots, y_n)$ space, then at least one integral curve of the system* (4.4) *passes through each interior point $(x_0, y_1^0, \ldots, y_n^0)$ of G.*

Proof. Just as in Sec. 11, we construct Euler lines L_k passing through $(x_0, y_1^0, \ldots, y_n^0)$ and then pass to the limit $|L_k| \to 0$, using Arzelà's theorem.[4]

---

[4] For the meaning of $|L_k|$, see p. 37. Also recall Remark 3, p. 32.

THEOREM 2 (*Osgood's uniqueness theorem*). *Suppose the functions* $f_i(x, y_1, \ldots, y_n)$ *satisfy the conditions*

$$|f_i(x, \tilde{y}_1, \ldots, \tilde{y}_n) - f_i(x, y_1^*, \ldots, y_n^*)| \leqslant \varphi\left(\sum_{k=1}^{n} |\tilde{y}_k - y_k^*|\right),$$
$$i = 1, \ldots, n \tag{4.11}$$

for every pair of points $(x, \tilde{y}_1, \ldots, \tilde{y}_n)$, $(x, y_1^*, \ldots, y_n^*)$ *in a domain* G, *where* $\varphi(u) > 0$ *is a continuous function on* $0 < u \leqslant a$ *and*

$$\lim_{\varepsilon \to 0+} \int_{\varepsilon}^{a} \frac{du}{\varphi(u)} = \infty.$$

Then there is no more than one integral curve of the system (4.4) *passing through each point* $(x_0, y_1^0, \ldots, y_n^0)$ *of* G.

Proof. The most common choice is $\varphi(u) = Ku$. In this case, (4.11) becomes

$$|f_i(x, \tilde{y}_1, \ldots, \tilde{y}_n) - f_i(x, y_1^*, \ldots, y_n^*)| \leqslant K \sum_{k=1}^{n} |\tilde{y}_k - y_k^*|,$$
$$i = 1, \ldots, n, \tag{4.12}$$

and is known as a *Lipschitz condition*.[5] Since the proof of Osgood's theorem is somewhat more complicated for a system of differential equations than for a single equation, we now supply the details.

Suppose there are two distinct solutions $\tilde{y}_1(x), \ldots, \tilde{y}_n(x)$ and $y_1^*(x), \ldots, y_n^*(x)$ of the system (4.4) such that

$$\tilde{y}_i(x_0) = y_i^*(x_0), \qquad i = 1, \ldots, n. \tag{4.13}$$

Since the solutions are distinct, there is a number x_1 such that

$$\sum_{i=1}^{n} |y_i(x_1) - y_i^*(x_1)| > 0.$$

There is no loss of generality in assuming that $x_1 > x_0$, since otherwise we need only replace x by $-x$. Despite the fact that the functions $y_i(x)$ and $y_i^*(x)$, and hence the differences $y_i(x)$ and $y_i^*(x)$, have derivatives, the absolute values of the difference $y_i(x) - y_i^*(x)$ will fail to have derivatives at all points where

$$y_i(x) - y_i^*(x) = 0, \qquad \frac{d}{dx}[y_i(x) - y_i^*(x)] \neq 0.$$

This compels us to consider right-hand (or left-hand) derivatives at these points, instead of ordinary derivatives. As usual, by the right-hand

---

[5] As we shall see in Secs. 30 and 31, if the functions y_i are continuous in all their arguments and satisfy a Lipschitz condition of the form (4.12), then existence and uniqueness can be proved by the method of successive approximations, just as in the case of a single equation $y' = f(x, y)$.

and left-hand derivatives of the function $z(x)$ at the point h, we mean the quantities

$$D_r z(x) = \lim_{\substack{h \to 0 \\ h > 0}} \frac{z(x + h) - z(x)}{h}, \qquad D_l z(x) = \lim_{\substack{h \to 0 \\ h < 0}} \frac{z(x + h) - z(x)}{h},$$

respectively. When it does not matter which of these derivatives is used, we will drop the subscripts r and l. It is easy to see that if the derivative $z'(x)$ exists, then so do both $D_r |z(x)|$ and $D_l |z(x)|$, and moreover

$$|D_r |z(x)| \,| = |D_l |z(x)| \,| = |z'(x)|,$$

or more concisely

$$|D |z(x)| \,| = |z'(x)|.$$

Taking all this into account, we use (4.11) to deduce from the identities

$$\frac{d\tilde{y}_i(x)}{dx} = f_i[x, \tilde{y}_1(x), \ldots, \tilde{y}_n(x)], \qquad i = 1, \ldots, n,$$

$$\frac{dy_i^*(x)}{dx} = f_i^*[x, y_1^*(x), \ldots, y_n^*(x)], \qquad i = 1, \ldots, n$$

that

$$|D| \,\tilde{y}_i(x) - y_i^*(x)| \,| = |f_i[x, \tilde{y}_1(x), \ldots, \tilde{y}_n(x)] - f_i^*[x, y_1^*(x), \ldots, y_n^*(x)]|$$

$$\leqslant \varphi\left(\sum_{k=1}^{n} |\tilde{y}_k(x) - y_k^*(x)| \right),$$

and hence

$$\left| D \sum_{i=1}^{n} |\tilde{y}_i(x) - y_i^*(x)| \right| \leqslant n\varphi\left(\sum_{k=1}^{n} |\tilde{y}_k(x) - y_k^*(x)| \right)$$

$$< (n + 1)\varphi\left(\sum_{k=1}^{n} |\tilde{y}_k(x) - y_k^*(x)| \right). \tag{4.14}$$

The last step is possible only if

$$\sum_{k=1}^{n} |\tilde{y}_k(x) - y_k^*(x)| > 0,$$

which, in particular, is the case for $x = x_1$. Setting

$$\sum_{i=1}^{n} |\tilde{y}_i(x) - y_i^*(x)| = z(x),$$

we now construct the solution of the equation

$$\frac{dy}{dx} = (n + 1)\varphi(y)$$

satisfying the condition $y(x_1) = z(x_1) = z_1$. Such a solution exists and is unique (see Sec. 4), and its graph approaches the negative x-axis asymptotically without ever intersecting the x-axis. By construction, the curves $z(x)$ and $y(x)$ intersect at the point (x_1, z_1), and the inequality

$$|D_l z(x_1)| < (n + 1)\varphi(z_1) = (n + 1)\varphi[y(x_1)] = y'(x_1)$$

immediately implies the existence of an interval $(x_1 - \varepsilon, x_1)$, $\varepsilon > 0$ on which

$$z(x) > y(x).$$

But this inequality holds for every ε, if $0 < \varepsilon \leqslant x_1$, since otherwise we immediately arrive at a contradiction by choosing ε to be the largest value compatible with the inequality. In fact, we would then have

$$D_r z(x_2) \geqslant y'(x_2) = (n + 1)\varphi[y(x_2)] = (n + 1)\varphi[z(x_2)],$$

since $z(x) > y(x)$ to the right of the point x_2, while on the other hand, since $z(x_2) > 0$, (4.14) implies the contradictory inequality

$$Dz(x_2) < (n + 1)\varphi[z(x_2)].$$

It follows that

$$z(x) \geqslant y(x) > 0$$

for all x in the interval $[x_0, x_1]$. In particular, $z(x_0) > 0$ contrary to (4.13). Thus there cannot be two distinct solutions of the system (4.4), and the proof is complete.

THEOREM 3 (*Implications for a system of higher-order equations*). *Given a system of differential equations*

$$\frac{d^{m_i}y_i}{dx^{m_i}} = f_i\left(x, y_1, \ldots, \frac{d^{m_1-1}y_1}{dx^{m_1-1}}, \ldots, y_n, \ldots, \frac{d^{m_n-1}y_n}{dx^{m_n-1}}\right), \tag{4.15}$$
$$i = 1, \ldots, n,$$

solved for the highest-order derivatives of the unknown functions $y_1, \ldots,$ *y_n, suppose every f_i is continuous in some neighborhood $\mathcal{N}$ of the point*

$$\left(x_0, y_1^0, \ldots, \left(\frac{d^{m_1-1}y_1}{dx^{m_1-1}}\right)^0, \ldots, y_n^0, \ldots, \left(\frac{d^{m_n-1}y_n}{dx^{m_n-1}}\right)^0\right)$$

and satisfies a Lipschitz condition in all its arguments except the first on $\mathcal{N}$. Then on some interval $[a, b]$ containing x_0 as an interior point there exists one and only one solution $y_1(x), \ldots, y_n(x)$ of the system (4.15) such that the functions

$$y_1(x), \ldots, \frac{d^{m_1-1}y_1(x)}{dx^{m_1-1}}, \ldots, y_n(x), \ldots, \frac{d^{m_n-1}y_n(x)}{dx^{m_n-1}}$$

take the values

$$y_1^0, \ldots, \left(\frac{d^{m_1-1}y_1}{dx^{m_1-1}}\right)^0, \ldots, y_n^0, \ldots, \left(\frac{d^{m_n-1}y_n}{dx^{m_n-1}}\right)^0$$

at the point $x = x_0$.

Proof. Use Theorems 1 and 2 and the considerations of Sec. 27. Just as in Sec. 11, the solution on $[a, b]$ can be continued in either direction.

THEOREM 4 (*Cauchy's theorem*). *If every function* $f_i(x, y_1, \ldots, y_n)$, $i = 1, \ldots, n$ *is analytic in all its arguments in a neighborhood of the point* $(x_0, y_1^0, \ldots, y_n^0)$, *then the system* (4.4) *has a unique solution* $y_1(x), \ldots, y_n(x)$ *satisfying the initial conditions*

$$y_1(x_0) = y_1^0, \ldots, y_n(x_0) = y_n^0,$$

and this solution is analytic in a neighborhood of x_0.

COROLLARY. *If every function* $f_i(x, y_1, \ldots, y_n)$, $i = 1, \ldots, n$ *is analytic in all its arguments on a domain G and if every* $f_i(x, y_1, \ldots, y_n)$ *is real for all real* $x, y_1, \ldots, y_n$, *then every real solution of the system* (4.4) *lying in G is analytic.*

THEOREM 5 (*Smoothness of solutions*). *If every function* $f_i(x, y_1, \ldots, y_n)$, $i = 1, \ldots, n$ *has continuous derivatives with respect to* $x, y_1, \ldots, y_n$ *up to order* $p \geqslant 0$ *(inclusive), then every solution of the system* (4.4) *has continuous derivatives with respect to* x *up to order* $p + 1$.

THEOREM 6 (*Dependence of the solution on parameters*). *Given a system*

$$\frac{dy_i}{dx} = f_i(x, y_1, \ldots, y_n, \mu_1, \ldots, \mu_m), \qquad i = 1, \ldots, n, \qquad (4.16)$$

let $\bar{G}_{xy_1 \cdots y_n}$ *be a closed domain in* $(x, y_1, \ldots, y_n)$ *space, let P be the open m-dimensional parallelepiped*

$$|\mu_1| < \mu_1^0, \ldots, |\mu_m| < \mu_m^0$$

(where the μ_i^0 *are positive constants), and let* $G = \bar{G}_{xy_1 \cdots y_n} \times P$. *Suppose every* $f_i(x, y_1, \ldots, y_n, \mu_1, \ldots, \mu_n)$ *has bounded continuous derivatives with respect to* $y_1, \ldots, y_n, \mu_1, \ldots, \mu_m$ *up to order* $p \geqslant 0$ *(inclusive) on G, and satisfies a Lipschitz condition in the variables* $y_1, \ldots, y_n$ *on G whose constant is independent of the parameters* $\mu_1, \ldots, \mu_m$. *Then, given any interior point* $(x, y_1^0, \ldots, y_n^0)$ *of* $\bar{G}_{xy_1 \cdots y_n}$, *there is a closed interval* $a \leqslant x \leqslant b$ $(a < x_0 < b)$ *on which the system* (4.4) *has a unique solution* $y_1 = \varphi_1(x, \mu_1, \ldots, \mu_m), \ldots, y_n = \varphi_n(x, \mu_1, \ldots, \mu_m)$ *taking the values* $y_1^0, \ldots, y_n^0$ *for* $x = x_0$. *Moreover, every* $\varphi_i(x, \mu_1, \ldots, \mu_m)$ *has continuous derivatives with respect to* $\mu_1, \ldots, \mu_m$ *up to order* p *on the set* $[a, b] \times P$.

COROLLARY. *If every* $f_i(x, y_1, \ldots, y_n)$ *has continuous derivatives with respect to* $x, y_1, \ldots, y_n$ *up to order* $p \geqslant 1$, *then the functions* $y_i(x, x_0, y_1^0, \ldots, y_n^0)$ *satisfying the system* (4.4) *and equal to* $y_1^0, \ldots, y_n^0$ *for* $x = x_0$

*have continuous derivatives with respect to $x_0, y_1^0, \ldots, y_n^0$ up to order $p \geqslant 1$.
This assertion remains true for $p = 0$ if the f_i are such as to guarantee
uniqueness of the solution passing through $(x_0, y_1^0, \ldots, y_n^0)$.*[6]

Problem 1. Verify Theorems 1 and 4 in detail, and also Theorem 6 and its
corollary. State the implications for systems of higher-order equations. Generalize the results of Secs. 13 and 19, and also Remark 3, p. 32, to the case of
systems. Also generalize Probs. 8 and 11 of Sec. 12, Prob. 4 of Sec. 14 and Prob.
5 of Sec. 19.

Problem 2. Prove that the assertion of Prob. 8, Sec. 19 generalizes to
surfaces of the form $y = f(x_1, \ldots, x_{n-1})$, but not to curves in n-dimensional
space $(n \geqslant 3)$.

Problem 3. Let every $f_i(x, y_1, \ldots, y_n)$ in the system (4.4) be continuous, and
suppose every solution of (4.4) satisfying the initial data stays inside the domain
G when continued onto the interval $a \leqslant x \leqslant b$ $(a < x_0 < b)$. Under these
conditions, prove that if the initial conditions and the functions f_i are subjected
to "sufficiently small" changes, then the varied solution can still be continued
onto the whole interval $[a, b]$, where it is uniformly close to one of the solutions
of (4.4) satisfying the original initial data.

Problem 4 (H. Kneser). Prove that under the conditions of Prob. 3, the
intersection of the plane $x = c$ $(a \leqslant c \leqslant b)$ with the "integral funnel" (the set
of all integral curves satisfying the given initial conditions) is nonempty, closed,
bounded and connected (i.e., cannot be represented as the union of two nonempty closed disjoint sets). For $n = 2$ construct an example where the indicated
intersection is a) a disk; b) a circle.

Hint. Consider the intersection of the plane $x = c$ with the set of all
Euler lines for systems with "neighboring" right-hand sides satisfying the given
initial conditions, where the projections of the segments of the Euler lines onto
the x-axis all have the same length.

Problem 5. Generalize Prob. 4 of Sec. 11 to the case of systems of the form
(4.4), and interpret the result geometrically (instead of a direction field, we have
a field of "cones of directions," where the cones are of a special form). Go from
cones of a special form to arbitrary convex cones. Generalize the results of the
preceding problems to systems of this kind.

Problem 6. Give sufficient conditions for existence and uniqueness of
solutions of an infinite system of equations

$$\frac{dy_i}{dx} = f_i(x, y_1, y_2, \ldots), \qquad i = 1, 2, \ldots,$$

involving infinitely many unknown functions.

---

[6] See the theorem of Sec. 19 and the remark on p. 58.

30. The Principle of Contraction Mappings for a Family of Vector Functions

An ordered array

$$\varphi(x) = (\varphi_1(x), \ldots, \varphi_n(x))$$

of n functions $\varphi_1(x), \ldots, \varphi_n(x)$ is called a *vector function* (in n dimensions),[7] and the separate functions $\varphi_1(x), \ldots, \varphi_n(x)$ are called the *components* of $\varphi(x)$. By $\varphi(x) \pm \tilde{\varphi}(x)$ we mean the vector function with components $\varphi_1(x) \pm \tilde{\varphi}_1(x), \ldots, \varphi_n(x) \pm \tilde{\varphi}_n(x)$. A sequence of vector functions

$$\varphi^{(k)}(x) = (\varphi_1^{(k)}(x), \ldots, \varphi_n^{(k)}(x)), \qquad k = 1, 2, \ldots$$

is said to *converge (uniformly)* as $k \to \infty$ to a vector function

$$\varphi(x) = (\varphi_1(x), \ldots, \varphi_n(x))$$

if each sequence of components $\varphi_i^{(k)}(x)$ converges (uniformly) as $k \to \infty$ to $\varphi_i(x)$, $i = 1, \ldots, n$. Clearly, *a necessary and sufficient condition for a sequence of vector functions $\varphi^{(k)}(x)$ to converge uniformly to $\varphi(x)$ as $k \to \infty$ is that the numerical sequence*

$$\sum_{i=1}^{n} \sup_{x \in E} |\varphi_i^{(k)}(x) - \varphi_i(x)|$$

converge to zero as $k \to \infty$ (why?).

A vector with only one component is often called a *scalar*. We now generalize the principle of contraction mappings proved in Sec. 15 for a family Φ of scalar functions:

THEOREM (*Principle of contraction mappings for a family of vector functions*).   *Let Φ be a nonempty family of vector functions $\varphi(x)$, each defined on the same set E, satisfying the following conditions:*

1) *Every function φ is bounded, in the sense that*

$$\sum_{i=1}^{n} \sup_{x \in E} |\varphi_i(x)| = M_\varphi < \infty$$

(in general, the bound depends on φ);

2) *The limit of any uniformly convergent sequence of functions in Φ also belongs to Φ;*

3) *There is an operator A defined on Φ carrying every function φ in Φ into another function $A\varphi = (\{A\varphi\}_1, \ldots, \{A\varphi\}_n)$ in Φ;*

---

[7] Later on, we shall deal with vector functions which are best thought of as "column vectors" rather than "row vectors."

4) *The inequality*

$$\sum_{i=1}^{n} \sup_{x \in E} |\{A\varphi(x)\}_i - \{A\tilde{\varphi}(x)\}_i| \leqslant m \sum_{i=1}^{n} \sup_{x \in E} |\varphi_i(x) - \tilde{\varphi}_i(x)| \quad (4.17)$$

holds for every pair of functions φ, $\tilde{\varphi}$ *in* Φ *and for some fixed m in the interval* $0 \leqslant m < 1$.

Then the equation

$$\varphi = A\varphi,$$

has one and only one solution in the family Φ.

Proof. The proof is word for word the same as in the scalar case (see Sec. 15) if we make the substitutions

$$\varphi_k \to \varphi_i^{(k)}, \qquad \sup |\varphi_{k+1} - \varphi_k| \to \sum_{i=1}^{n} \sup |\varphi_i^{(k+1)} - \varphi_i^{(k)}|,$$

$$\sup |A\varphi - A\varphi_{k-1}| \to \sum_{i=1}^{n} \sup |\{A\varphi\}_i - \{A\varphi^{(k-1)}\}_i|,$$

and so on, bearing in mind the italicized assertion preceding the statement of the theorem.

Remark. The geometric interpretation of the theorem is the same as in the scalar case (see Sec. 16), except that now the "points" are vector functions (curves) and the "distance" between two points $\varphi = (\varphi_1, \ldots, \varphi_n)$ and $\tilde{\varphi} = (\tilde{\varphi}_1, \ldots, \tilde{\varphi}_n)$ is given by

$$\sum_{i=1}^{n} \sup |\varphi_i - \tilde{\varphi}_i|.$$

Problem. Prove that the principle of contraction mappings remains true if the inequality (4.17) is generalized to

$$F(\sup |\{A\varphi\}_1 - \{A\tilde{\varphi}\}_1|, \ldots, \sup |\{A\varphi\}_n - \{A\tilde{\varphi}\}_n|)$$
$$\leqslant mF(\sup |\varphi_1 - \tilde{\varphi}_1|, \ldots, \sup |\varphi_n - \tilde{\varphi}_n|),$$

where $F(t_1, \ldots, t_n)$ is a continuous nonnegative homogeneous function of degree 1, defined for $t_1 \geqslant 0, \ldots, t_n \geqslant 0$ and zero only at the origin. Give examples of such functions.

31. The Method of Successive Approximations for a System of Differential Equations

THEOREM. *Suppose the functions* $f_i(x, y_1, \ldots, y_n)$, $i = 1, \ldots, n$ *appearing in* (4.4) *are continuous in x on a domain G in* $(x, y_1, \ldots, y_n)$ *space and satisfy a Lipschitz condition in the variables* $y_1, \ldots, y_n$ *on*

every closed bounded domain $\bar{G}'$ contained in G. Then, given any point
$(x_0, y_1^0, \ldots, y_n^0)$ in G, there is a closed interval $[a, b]$ containing x_0 as an
interior point on which the system (4.4) has a unique solution $y_1(x), \ldots,$
$y_n(x)$ taking the values $y_1^0, \ldots, y_n^0$ for $x = x_0$.

Proof. First we note that if such a solution $y_1(x), \ldots, y_n(x)$
exists, the integration of the identities

$$\frac{dy_i(\xi)}{d\xi} = f_i[\xi, y_1(\xi), \ldots, y_n(\xi)], \qquad i = 1, \ldots, n$$

between x and x_0 gives a system of integral equations[8]

$$y_i(x) = y_i^0 + \int_{x_0}^{x} f_i[\xi, y_1(\xi), \ldots, y_n(\xi)] \, d\xi, \qquad i = 1, \ldots, n. \quad (4.18)$$

Thus every solution of the system (4.4) passing through the point
$(x_0, y_1^0, \ldots, y_n^0)$ satisfies (4.18). Conversely, every continuous solution
$y_1(x), \ldots, y_n(x)$ of the system of integral equations (4.18) satisfies
the system of differential equations (4.4) and the initial conditions[9]

$$y_1(x_0) = y_1^0, \ldots, y_n(x_0) = y_n^0.$$

In fact, the initial conditions are obviously satisfied by the left-hand
sides of (4.18), and moreover differentiation of both sides of each
equation (4.18) leads at once to the system (4.4), provided the differentiation is legitimate. But the right-hand side of each equation (4.18)
is clearly differentiable for continuous $y_1(x), \ldots, y_n(x)$ [being an indefinite integral of a continuous function], and hence so is the left-hand
side. Therefore, instead of proving that the system (4.4) has a unique
solution taking the values $y_1^0, \ldots, y_n^0$ for $x = x_0$ on some closed interval
$[a, b]$ $(a < x_0 < b)$, we shall now prove that the system of integral
equations (4.18) has a unique continuous solution on $[a, b]$. This
will be done by applying the principle of contraction mappings for
a family of vector functions.

With this in mind, let $\bar{G}'$ be any closed bounded domain contained
in G with $(x_0, y_1^0, \ldots, y_n^0)$ as an interior point, and let M be the least
upper bound of all the functions $|f_i(x, y_1, \ldots, y_n)|$, $i = 1, \ldots, n$ on
$\bar{G}'$. Draw $2n$ planes

$$y_i - y_i^0 = \pm M(x - x_0), \qquad i = 1, \ldots, n \quad (4.19)$$

through the point $(x_0, y_1^0, \ldots, y_n^0)$, and then two planes $x = a$ and $x = b$
which together with the planes (4.19) form two pyramids P_1 and P_2
completely contained in G', with common vertex at $(x_0, y_1^0, \ldots, y_n^0)$

---

[8] Each integrand is continuous by an argument like that in footnote 9, p. 39.
[9] See footnote 10, p. 39.

[later in the proof, we shall assume that a and b are sufficiently close to x_0]. Now let E be the interval $[a, b]$ and Φ the family of all continuous vector functions[10] $\varphi(x) = (\varphi_1(x), \ldots, \varphi_n(x))$ defined on $[a, b]$ whose graphs lie in $\bar{G}'$. Then conditions 1 and 2 of the principle of contraction mappings are obviously satisfied. Moreover, let A be the operator carrying φ into $A\varphi$, where

$$(A\varphi)_i = y_i^0 + \int_{x_0}^x f_i[\xi, \varphi_1(\xi), \ldots, \varphi_n(\xi)] \, d\xi, \qquad i = 1, \ldots, n.$$

It is clear that $\varphi^*(x) = A\varphi(x)$ is itself defined and continuous on $[a, b]$, and in addition satisfies the initial conditions $\varphi_1^*(x_0) = y_1^0, \ldots, \varphi_n^*(x_0) = y_n^0$. Moreover, the curve $y_i = \varphi_i^*(x)$, $i = 1, \ldots, n$ cannot leave the pyramids P_1 and P_2, and hence cannot leave $\bar{G}'$, since

$$|f_i[\xi, \varphi_1(\xi), \ldots, \varphi_n(\xi)]| \leqslant M, \qquad i = 1, \ldots, n$$

implies

$$|\varphi_i^*(x) - y_i^0| \leqslant M |x - x_0|, \qquad i = 1, \ldots, n.$$

Therefore condition 3 of the principle of contraction mappings is satisfied. To verify condition 4, we use the fact that every f_i satisfies a Lipschitz condition in the variables $y_1, \ldots, y_n$ to write

$$\sum_{i=1}^n \sup |\{A\varphi\}_i - \{A\tilde{\varphi}\}_i|$$

$$= \sum_{i=1}^n \sup \left| \left(y_i^0 + \int_{x_0}^x f_i[\xi, \varphi_1(\xi), \ldots, \varphi_n(\xi)] \, d\xi \right) \right.$$
$$\left. - \left(y_i^0 + \int_{x_0}^x f_i[\xi, \tilde{\varphi}_1(\xi), \ldots, \tilde{\varphi}_n(\xi)] \, d\xi \right) \right|$$

$$= \sum_{i=1}^n \sup \left| \int_{x_0}^x \{f_i[\xi, \varphi_1(\xi), \ldots, \varphi_n(\xi)] - f_i[\xi, \tilde{\varphi}_1(\xi), \ldots, \tilde{\varphi}_n(\xi)]\} \, d\xi \right|$$

$$\leqslant K \sum_{i=1}^n \sup \int_{x_0}^x [|\varphi_1(\xi) - \tilde{\varphi}_1(\xi)| + \cdots + |\varphi_n(\xi) - \tilde{\varphi}_n(\xi)|] \, d\xi$$

$$\leqslant Kn(b - a) \sum_{i=1}^n \sup |\varphi_i - \tilde{\varphi}_i| = m \sum_{i=1}^n \sup |\varphi_i - \tilde{\varphi}_i|,$$

where $Kn(b - a) = m$ and $m < 1$ if $[a, b]$ is sufficiently small. Therefore we can use the principle of contraction mappings to deduce the existence of a unique continuous solution of the system of integral equations (4.18), or equivalently the existence of a unique solution $y_1(x), \ldots, y_n(x)$ of the system of differential equations (4.4) satisfying the initial conditions $y_1(x_0) = y_1^0, \ldots, y_n(x_0) = y_n^0$.

---

[10] A vector function is said to be continuous if all its components are continuous.

Remark 1. All the remarks made at the end of Sec. 11 remain in force here.

Remark 2. Suppose the domain G contains a "strip" $c \leqslant x \leqslant d, -\infty < y_i < \infty, i = 1, \ldots, n$ (where c and d are finite), and suppose the functions f_i satisfy a uniform Lipschitz condition in the strip. Then, just as in Remark 3, p. 42, it can be shown that if the "zeroth approximation" (i.e., the function φ_0 of Secs. 14 and 15) is continuous on $[a, b]$, the successive approximations converge uniformly on $[c, d]$ to a solution of the system (4.4).

Instead of studying singular points, singular curves and more generally, singular surfaces for systems of differential equations (as in Chap. 3), we now turn our attention to systems of *linear* differential equations.

5

LINEAR SYSTEMS: GENERAL THEORY

---

32. Definitions. Implications of the General Theory of Systems

A system of differential equations is said to be *linear* if every equation of the system is linear in the unknown functions and their derivatives. As shown in Sec. 27, every system of differential equations is equivalent to a system containing only first derivatives. Therefore we shall deal primarily with *first-order* linear systems, confining ourselves as before to the case of equations which are solved with respect to the derivatives. The general form of such a system is

$$\frac{dy_i}{dx} = \sum_{j=1}^{n} a_{ij}(x)y_j + f_i(x), \qquad i = 1, \ldots, n, \tag{5.1}$$

which can be greatly simplified by using matrix notation. In fact, suppose we introduce the $n \times n$ coefficient matrix

$$A(x) = \|a_{ij}(x)\|$$

and the $n \times 1$ column matrices (or vectors)

$$f(x) = \begin{Vmatrix} f_1(x) \\ f_2(x) \\ \cdot \\ \cdot \\ \cdot \\ f_n(x) \end{Vmatrix}, \qquad y(x) = \begin{Vmatrix} y_1(x) \\ y_2(x) \\ \cdot \\ \cdot \\ \cdot \\ y_n(x) \end{Vmatrix}$$

(cf. footnote 7, p. 98), which are handled by the usual rules familiar from linear algebra. (In particular, the derivative of a matrix or a vector is found simply by differentiating its components.) Then (5.1) takes the form

$$\frac{dy}{dx} = A(x)y + f(x), \tag{5.2}$$

as can be verified by calculating the matrix product in the right-hand side.

From now on we shall assume that the functions $a_{ij}(x)$ and $f_i(x)$ are *continuous* in an interval $a < x < b$, where this interval may become infinite at either end (or both). The right-hand sides of such systems have bounded derivatives with respect to every y_i and hence satisfy Lipschitz conditions in every closed interval $[a_1, b_1]$ contained in (a, b). Therefore it follows from the theorem of Sec. 31 that one and only one integral curve of the system (5.1) passes through every point $(x_0, y_1^0, \ldots, y_n^0)$ of the strip $a < x < b$ in $(x, y_1, \ldots, y_n)$ space. In fact, the theorem is immediately applicable to every parallelepiped of the form

$$a + \varepsilon \leqslant x \leqslant b - \varepsilon, \quad -M \leqslant y_i \leqslant M, \qquad i = 1, \ldots, n,$$

where $\varepsilon > 0$ is arbitrarily small and $M > 0$ arbitrarily large. But then the theorem also holds in the entire strip $a < x < b$.

Every solution of the system (5.1) *can be extended onto the whole interval* (a, b). This fact follows at once from the next to the last paragraph of Sec. 31, as applied to the closed interval $[a_1, b_1]$ contained in (a, b). Thus the functions $y_1(x), \ldots, y_n(x)$ constituting the solution of the system (5.1) may go to infinity as $x \to a$ or $x \to b$.

If every $f_i(x) \equiv 0$, we call the system (5.1) *homogeneous* [in this case, $f(x) \equiv 0$ in equation (5.2)]. Otherwise the system is called *nonhomogeneous*.

Problem 1. By the *norm* $\|A\|$ of a square or rectangular matrix is meant the sum of the absolute values of its elements. Prove that

$$\|A + B\| \leqslant \|A\| + \|B\|,$$
$$\|cA\| = |c| \, \|A\| \quad \text{(c a number)},$$
$$\|AB\| \leqslant \|A\| \, \|B\|.$$

Prove that if the elements of a matrix A depend on x and have a right-hand (left-hand) derivative, then $\|A(x)\|$ has a right-hand (left-hand) derivative $D_r \|A(x)\|$, where

$$|D_r \|A(x)\| \, | \leqslant \|D_r A(x)\|.$$

Problem 2. Using the preceding problem, estimate $|D_r \|y(x)\| \, |$ where $y(x)$ is the vector solution of equation (5.2). Using this result and Prob. 7, Sec. 12, estimate the rate of possible growth of the solution as $x \to a$ or $x \to b$. For example,

$$\|y(x)\| \leqslant \|y(x_0)\| \exp \left\{ \int_{x_0}^{x} \|A(\xi)\| \, d\xi \right\} + \int_{x_0}^{x} \|f(s)\| \exp \left\{ \int_{s}^{x} \|A(\xi)\| \, d\xi \right\} ds.$$

Problem 3. Prove that if all the functions $a_{ij}(x)$ and $f_i(x)$ can be expanded in Maclaurin series with radius of convergence no less than $R > 0$, then every solution of the system (5.1) can also be expanded in a Maclaurin series with radius of convergence no less than R. This is proved in the same way as the analogous assertion in the remark on p. 54. Choose the same majorizing function for every $a_{ij}(x)$ and $f_i(x)$.

33. Basic Theorems for Homogeneous First-Order Systems

By a *linear combination* of m solutions

$$y^{(1)}(x) = \begin{Vmatrix} y_1^{(1)}(x) \\ y_2^{(1)}(x) \\ \cdot \\ \cdot \\ \cdot \\ y_n^{(1)}(x) \end{Vmatrix}, \ldots, y^{(m)}(x) = \begin{Vmatrix} y_1^{(m)}(x) \\ y_2^{(m)}(x) \\ \cdot \\ \cdot \\ \cdot \\ y_n^{(m)}(x) \end{Vmatrix} \qquad (5.3)$$

of a homogeneous linear system, we mean a vector function of the form

$$\sum_{k=1}^{m} C_k y^{(k)}(x),$$

where $C_1, \ldots, C_m$ are constants. In particular, if every $C_k = 1$, we speak of the *sum* of the solutions, while if $C_1 = 1$, $C_2 = -1$, $m = 2$, we speak of the *difference* between the solutions $y^{(1)}(x)$ and $y^{(2)}(x)$.

THEOREM 1. *A linear combination of solutions of a homogeneous linear system is also a solution of the system.*

Proof. Let $y^{(1)}(x), \ldots, y^{(m)}(x)$ be m solutions of the homogeneous linear system

$$\frac{dy}{dx} - A(x)y = 0 \qquad (5.4)$$

or

$$\frac{dy_i}{dx} - \sum_{j=1}^{n} a_{ij} y_j = 0, \qquad i = 1, \ldots, n. \qquad (5.4')$$

If y is replaced by

$$\sum_{k=1}^{m} C_k y^{(k)}(x)$$

in (5.4), the left-hand side becomes

$$\frac{d}{dx} \sum_{k=1}^{m} C_k y^{(k)}(x) - A(x) \sum_{k=1}^{m} C_k y^{(k)}(x) = \sum_{k=1}^{m} \left[C_k \frac{dy^{(k)}(x)}{dx} - A(x)y^{(k)}(x) \right].$$

But every function (5.3) satisfies (5.4), and hence

$$\frac{dy^{(k)}(x)}{dx} - A(x)y^{(k)}(x) = 0, \qquad k = 1, \ldots, m.$$

It follows that

$$\frac{d}{dx} \sum_{k=1}^{m} C_k y^{(k)}(x) - A(x) \sum_{k=1}^{m} C_k y^{(k)}(x) \equiv 0,$$

as required.

DEFINITION. *The m vector functions (5.3) are said to be linearly dependent if there exist constants $C_1, \ldots, C_m$, not all zero, such that*

$$\sum_{k=1}^{m} C_k y^{(k)}(x) \equiv 0.$$

Otherwise, they are said to be linearly independent. The determinant

$$W(x) = \begin{vmatrix} y_1^{(1)}(x) & y_1^{(2)}(x) & \cdots & y_1^{(n)}(x) \\ y_2^{(1)}(x) & y_2^{(2)}(x) & \cdots & y_2^{(n)}(x) \\ \cdot \cdot \cdot \cdot \cdot \cdot \cdot \cdot \cdot \cdot \cdot \cdot \cdot \cdot \\ y_n^{(1)}(x) & y_n^{(2)}(x) & \cdots & y_n^{(n)}(x) \end{vmatrix}$$

is called the Wronskian of the set of vector functions

$$y^{(1)}(x) = \begin{Vmatrix} y_1^{(1)}(x) \\ y_2^{(1)}(x) \\ \cdot \\ \cdot \\ \cdot \\ y_n^{(1)}(x) \end{Vmatrix}, \ldots, y^{(n)}(x) = \begin{Vmatrix} y_1^{(n)}(x) \\ y_2^{(n)}(x) \\ \cdot \\ \cdot \\ \cdot \\ y_n^{(n)}(x) \end{Vmatrix}. \tag{5.5}$$

THEOREM 2. *If the functions (5.5) are linearly dependent, then their Wronskian vanishes identically.*

Proof. This result is an immediate consequence of a familiar theorem of linear algebra.

THEOREM 3. *Suppose the functions (5.5), with Wronskian $W(x)$, are solutions of the system (5.4). Then the functions are linearly dependent if $W(x)$ vanishes at any point $x = x_0$.*

Proof. If $W(x_0) = 0$, the vectors $y^{(1)}(x_0), \ldots, y^{(n)}(x_0)$ are linearly dependent, and hence we can find constants $C_1^*, \ldots, C_n^*$, not all zero, such that

$$\sum_{k=1}^{n} C_k^* y^{(k)}(x_0) = 0.$$

Now consider the vector function

$$y^*(x) = \sum_{k=1}^{n} C_k^* y^{(k)}(x).$$

This function satisfies the system (5.4), by Theorem 1, and equals the zero vector. But by the uniqueness theorem, there can be only one function which satisfies the system (5.4) and vanishes for $x = x_0$, and this function is obviously the function which vanishes identically. Therefore

$$\sum_{k=1}^{n} C_k^* y^{(k)}(x) \equiv 0,$$

and the theorem is proved.

COROLLARY. *If the Wronskian formed from the solutions (5.5) of the system (5.4) vanishes at a single point, then it vanishes identically.*

Proof. By Theorem 3, the functions (5.5) are linearly dependent, and by Theorem 2, their Wronskian vanishes identically.

Remark. Theorem 3 breaks down if the functions (5.5) are not solutions of a system of the form (5.4) with continuous coefficients. For example, the Wronskian of the functions

$$\left\| \begin{array}{c} x \\ 0 \end{array} \right\|, \quad \left\| \begin{array}{c} x^2 \\ 0 \end{array} \right\|$$

vanishes identically, although they are linearly independent.

DEFINITON. *A set of n linearly independent solutions of the system (5.4) is called a fundamental set of solutions of (5.4).*

THEOREM 4. *Fundamental sets of solutions exist.*

Proof. Choose n^2 numbers $b_i^{(k)}$ such that

$$\begin{vmatrix} b_1^{(1)} & b_1^{(2)} & \cdots & b_1^{(n)} \\ b_2^{(1)} & b_2^{(2)} & \cdots & b_2^{(n)} \\ \cdot & \cdot & \cdots & \cdot \\ b_n^{(1)} & b_n^{(2)} & \cdots & b_n^{(n)} \end{vmatrix} \neq 0,$$

e.g., the numbers

$$b_i^{(k)} = \begin{cases} 0 & \text{for} \quad i \neq k, \\ 1 & \text{for} \quad i = k. \end{cases}$$

Now form $m = n$ solutions (5.3) of the system (5.4) satisfying the conditions

$$y_i^{(k)}(x_0) = b_i^{(k)}, \qquad i, k = 1, \ldots, n,$$

where x_0 is any point in the interval (a, b). Then the Wronskian of these solutions is nonzero for $x = x_0$, and hence the solutions are linearly independent, by Theorem 2.

THEOREM 5.   *If the functions* (5.5) *are n linearly independent solutions of the homogeneous linear system* (5.4), *then every solution $y(x)$ of* (5.4) *can be represented as a linear combination of these solutions with suitably chosen constants i.e.,*

$$y(x) = \sum_{k=1}^{n} C_k y^{(k)}(x).$$

Equivalently, the general solution of (5.4) *is a linear combination with arbitrary coefficients of any fundamental set of solutions.*

Proof.   Let the vector function $y(x)$ be any solution of (5.4), taking the value $y(x_0)$ for $x = x_0$. Since $W(x_0) \neq 0$, the vectors $y^{(1)}(x_0), \ldots, y^{(n)}(x_0)$ are linearly independent, and hence $y(x_0)$ can be represented as a linear combination of $y^{(1)}(x_0), \ldots, y^{(n)}(x_0)$:

$$y(x_0) = \sum_{k=1}^{n} C_k^* y^{(k)}(x_0). \tag{5.6}$$

Now consider the vector function

$$y^*(x) = \sum_{k=1}^{n} C_k^* y^{(k)}(x).$$

By Theorem 1, $y^*(x)$ satisfies the system (5.4). But on the other hand, $y^*(x)$ takes the same value as $y(x)$ for $x = x_0$. Therefore, by the uniqueness theorem, $y(x) \equiv y^*(x)$, i.e.,

$$y(x) \equiv \sum_{k=1}^{n} C_k^* y^{(k)}(x),$$

and the theorem is proved.

Problem 1.   Let $Y(x)$ be a square matrix whose columns consist of any n solutions of the system (5.4). Show that such a matrix satisfies the equation

$$\frac{dY}{dx} = A(x) Y,$$

where either det $Y(x) \equiv 0$ or det $Y(x) \neq 0$. In the latter case, $Y(x)$ is called a *fundamental matrix* of the system (5.4). Show that

  a) If $Y(x)$ is a fundamental matrix, then the general solution of (5.4) is of the form $y = Y(x)C$, where C is any constant vector;

  b) The solution of (5.4) satisfying the initial condition $y(x_0) = y^0$ is given by $y = Y(x)[Y(x_0)]^{-1}y^0$.

Problem 2. Find all the solutions of the system

$$xy_1' = 2y_1 - y_2,$$
$$xy_2' = 2y_1 - y_2.$$

Show that

    a) If the initial conditions are specified for $x_0 \neq 0$, the solution exists and is unique on the entire real axis, but if $x_0 = 0$, the solution exists only if $2y_1^0 - y_2^0 = 0$ and is not unique;

    b) The Wronskian of every pair of linearly independent solutions equals cx, where $c \neq 0$.

How can the fact that the Wronskian vanishes at only one point be reconciled with the corollary to Theorem 3?

Problem 3. Solve the system

$$xy_1' = y_1 - 2y_2,$$
$$xy_2' = y_1 - 2y_2.$$

Show that a solution exists on the entire real axis if and only if the initial data satisfies the condition $y_1^0 = 2y_2^0$. Show that the solution is then unique.

Problem 4. Find a fundamental set of solutions and its Wronskian for the system

$$y_i' = \sum_{j=1}^{n} a_j(x)y_j, \qquad i = 1, \ldots, n,$$

where the functions $a_j(x)$ are all continuous on the interval (a, b).

34. An Expression for the Wronskian

    THEOREM. *Suppose the n functions* (5.5) *are solutions of the homogeneous linear system* (5.4), *with Wronskian* $W(x)$. *Then the relation between the values of the Wronskian at the points x and x_0 is given by the expression*[1]

$$W(x) = W(x_0) \exp\left\{ \int_{x_0}^{x} [a_{11}(\xi) + \cdots + a_{nn}(\xi)] \, d\xi \right\}. \tag{5.7}$$

    Proof. By the rule for differentiating a determinant, we have

$$W'(x) = \begin{vmatrix} \dfrac{dy_1^{(1)}}{dx} & \dfrac{dy_1^{(2)}}{dx} & \cdots & \dfrac{dy_1^{(n)}}{dx} \\ y_2^{(1)} & y_2^{(2)} & \cdots & y_2^{(n)} \\ \cdots & \cdots & \cdots & \cdots \\ y_n^{(1)} & y_n^{(2)} & \cdots & y_n^{(n)} \end{vmatrix} + \cdots + \begin{vmatrix} y_1^{(1)} & y_1^{(2)} & \cdots & y_1^{(n)} \\ y_2^{(1)} & y_2^{(2)} & \cdots & y_2^{(n)} \\ \cdots & \cdots & \cdots & \cdots \\ \dfrac{dy_n^{(1)}}{dx} & \dfrac{dy_n^{(2)}}{dx} & \cdots & \dfrac{dy_n^{(n)}}{dx} \end{vmatrix}.$$

---

[1] Found by Abel in 1827 for second-order equations, and by Liouville and Ostrogradski in 1838 for the general case.

Moreover, according to (5.4'),

$$\frac{dy_i^{(k)}}{dx} = \sum_{j=1}^{n} a_{ij}(x) y_j^{(k)}.$$

Substituting these expressions for the derivatives $dy_i^{(k)}/dx$ into the formula for $W'(x)$, and using the fact that the value of a determinant does not change if the elements of a row are multiplied by any number and then added to any other row, we find that

$$W'(x) = \begin{vmatrix} a_{11}y_1^{(1)} & a_{11}y_1^{(2)} & \cdots & a_{11}y_1^{(n)} \\ y_2^{(1)} & y_2^{(2)} & \cdots & y_2^{(n)} \\ \cdot & \cdot \cdot \cdot \cdot \cdot \cdot \cdot & \cdot & \cdot \\ y_n^{(1)} & y_n^{(2)} & \cdots & y_n^{(n)} \end{vmatrix}$$

$$+ \cdots + \begin{vmatrix} y_1^{(1)} & y_1^{(2)} & \cdots & y_1^{(n)} \\ y_2^{(1)} & y_2^{(2)} & \cdots & y_2^{(n)} \\ \cdot & \cdot \cdot \cdot \cdot \cdot \cdot \cdot \cdot \cdot & \cdot & \cdot \\ a_{nn}y_n^{(1)} & a_{nn}y_n^{(2)} & \cdots & a_{nn}y_n^{(n)} \end{vmatrix}$$

or

$$W'(x) = \sum_{i=1}^{n} a_{ii}(x) W(x).$$

Integration of this differential equation gives (5.7), as required.

Remark. This theorem immediately leads to another proof of the corollary to Theorem 3.

35. Formation of a Homogeneous Linear System from a Fundamental Set of Solutions

We begin by noting that a set of n vector functions $y^{(1)}, \ldots, y^{(n)}$ with continuous first derivatives may not be a fundamental set of solutions of a system of the form (5.4) with continuous coefficients. In fact, according to Theorem 3, p. 106, a necessary condition for this to be the case is that the Wronskian of the functions $y^{(1)}, \ldots, y^{(n)}$ vanish nowhere. We now show that this condition is also sufficient:

THEOREM. *Let $y^{(1)}, \ldots, y^{(n)}$ be n vector functions of the form (5.5) with continuous first derivatives, and suppose the Wronskian of $y^{(1)}, \ldots, y^{(n)}$*

is nonvanishing. Then $y^{(1)}, \ldots, y^{(n)}$ *is a fundamental set of solutions of a unique homogeneous linear system of the form* (5.4).

Proof. Consider the following system of homogeneous linear differential equations satisfied by n unknown functions $y_1(x), \ldots, y_n(x)$:

$$\begin{vmatrix} y_1 & y_1^{(1)} & \cdots & y_1^{(n)} \\ y_2 & y_2^{(1)} & \cdots & y_2^{(n)} \\ \cdot & \cdot & \cdot \cdot \cdot \cdot & \cdot \\ y_n & y_n^{(1)} & \cdots & y_n^{(n)} \\ \dfrac{dy_i}{dx} & \dfrac{dy_i^{(1)}}{dx} & \cdots & \dfrac{dy_i^{(n)}}{dx} \end{vmatrix} = 0, \qquad i = 1, \ldots, n.$$

It is easy to see that any of the functions (5.5) satisfies these equations. Moreover, these equations can all be solved for the derivatives dy_i/dx ($i = 1, \ldots, n$), since the Wronskian of the functions $y^{(1)}, \ldots, y^{(n)}$ is nonvanishing by hypothesis. To prove the uniqueness of the resulting system, we note that according to Theorem 5 of Sec. 33, every solution of the system is a linear combination of any fundamental set of solutions. But specifying all the integral curves of the system uniquely determines the direction field, and hence the right-hand sides of the equations

$$\frac{dy_i}{dx} = \sum_{j=1}^{n} a_{ij}(x) y_j,$$

since the coefficients of a linear form are uniquely determined by its values.

Problem. Given a set of m continuously differentiable functions of the form (5.3), where $m < n$, show that this set can be enlarged (by including new functions) to make a fundamental set of solutions (5.5) of some system (5.4) with continuous coefficients if and only if the rank of the matrix (5.3) equals m at every point of the interval (a, b).

Hint. First construct the system (5.4) in a neighborhood of any point of (a, b).

36. Implications for Equations of Order n

According to Sec. 27, the homogeneous linear differential equation

$$\frac{d^n y}{dx^n} = a_{n-1}(x) \frac{d^{n-1}y}{dx^{n-1}} + a_{n-2}(x) \frac{d^{n-2}y}{dx^{n-2}} + \cdots + a_1(x) \frac{dy}{dx} + a_0(x) y \quad (5.8)$$

is equivalent to the homogeneous linear system

$$\frac{dy_0}{dx} = y_1,$$

$$\frac{dy_1}{dx} = y_2,$$

$$\cdots \cdots \cdots \cdots \cdots \cdots \cdots \cdots \cdots \cdots \cdots \cdots \quad (5.9)$$

$$\frac{dy_{n-2}}{dx} = y_{n-1},$$

$$\frac{dy_{n-1}}{dx} = a_0(x)y_0 + a_1(x)y_1 + a_2(x)y_2 + \cdots + a_{n-1}(x)y_{n-1},$$

where y_0 denotes y and y_k denotes the kth derivative of y.

Concerning the equation (5.8), we now make the following observations:

1. According to Sec. 32, *if the coefficients $a_i(x)$ are continuous on the interval (a, b), then given any x_0 in (a, b) and any set of numbers $y_0^0, y_1^0, \ldots, y_{n-1}^0$, there exists one and only one solution of (5.8) whose derivative of order i $(i = 0, 1, \ldots, n - 1)$ equals y_i^0 at the point $x = x_0$. This solution exists on the whole interval (a, b).*

2. *It is obvious that if the functions $y^{(1)}(x), \ldots, y^{(m)}(x)$ satisfy (5.8), then so does any linear combination*

$$\sum_{k=1}^{m} C_k y^{(k)}(x) \quad (5.10)$$

with constant coefficients.[2]

3. Given m solutions

$$\left\| \begin{array}{c} y_0^{(k)}(x) \\ y_1^{(k)}(x) \\ \cdot \\ \cdot \\ \cdot \\ y_{n-1}^{(k)}(x) \end{array} \right\|, \qquad k = 1, \ldots, m$$

of the system (5.9), suppose there exist constants $C_1, \ldots, C_m$ such that

$$\sum_{k=1}^{m} C_k y_0^{(k)}(x) \equiv 0 \quad (5.11)$$

for all x in the interval (a, b). Then clearly we also have the identities

$$\sum_{k=1}^{m} C_k y_i^{(k)}(x) \equiv 0, \qquad i = 1, \ldots, n - 1.$$

---

[2] Here the functions $y^{(k)}(x)$ are scalar (ordinary) functions, and not vector functions as before.

Therefore we call the solutions $y^{(1)}(x), \ldots, y^{(m)}(x)$ of equation (5.8) *linearly dependent* if there exist constants $C_1, \ldots, C_m$, not all zero, such that

$$\sum_{k=1}^{m} C_k y^{(k)}(x) \equiv 0.$$

4. Bearing in mind that $y_i(x)$ is the derivative of order i of $y(x)$, we see that the appropriate definition of the Wronskian of n functions $y^{(1)}, \ldots, y^{(n)}$ is now

$$\begin{vmatrix} y^{(1)} & y^{(2)} & \cdots & y^{(n)} \\ \dfrac{dy^{(1)}}{dx} & \dfrac{dy^{(2)}}{dx} & \cdots & \dfrac{dy^{(n)}}{dx} \\ \cdot & \cdot \cdot \cdot \cdot \cdot \cdot \cdot & \cdot \cdot & \cdot \\ \dfrac{d^{n-1}y^{(1)}}{dx^{n-1}} & \dfrac{d^{n-1}y^{(2)}}{dx^{n-1}} & \cdots & \dfrac{d^{n-1}y^{(n)}}{dx^{n-1}} \end{vmatrix}. \qquad (5.12)$$

Thus Theorems 2, 3, 4 *and* 5 *of Sec.* 33 (*and also the corollary to Theorem* 3) *all continue to apply for the equation* (5.8), *provided we interpret a linear combination of solutions to be a sum of the form* (5.10) *and the Wronskian to be a determinant of the form* (5.12).

5. There is only one nonzero coefficient on the principal diagonal in the right-hand side of the system (5.9), namely the coefficient a_{n-1}. *Therefore formula* (5.7) *now takes the form*

$$W(x) = W(x_0) \exp \left\{ \int_{x_0}^{x} a_{n-1}(\xi)\, d\xi \right\}.$$

Remark. Suppose the determinant (5.12) vanishes identically, but the functions $y^{(1)}(x), \ldots, y^{(n)}(x)$ are not assumed to satisfy an equation of the form (5.8) with continuous coefficients. Then it cannot be concluded that the functions are linearly independent, i.e., that they satisfy an identity of the form (5.11) with $m = n$, where the coefficients C_k are not all zero. For example, the functions

$$y_1(x) = x^2, \qquad y_2(x) = x\,|x| \qquad (-1 \leqslant x \leqslant 1)$$

are linearly independent, although their Wronskian vanishes identically.

Problem 1. Show that if

$$y_{i+1}^{(k)} = \frac{d^i y^{(k)}}{dx^i}, \qquad i = 0, \ldots, n-1;\ k = 1, \ldots, n,$$

then the system constructed in Sec. 35 is equivalent to a single nth-order equation.

Problem 2. Suppose the solution $y(x)$ of equation (5.8) has infinitely many zeros in an interval $[a_1, b_1]$, where $a < a_1 < b_1 < b$. Prove that $y(x) \equiv 0$ on (a, b).

Problem 3. Solve the equation

$$y'' + xy = 0$$

by expanding y in a Maclaurin series. Establish the convergence of this series (cf. Prob. 3, Sec. 32).

Problem 4. Prove that if the Wronskian (5.12) vanishes identically for *analytic* functions $y^{(1)}(x), \ldots, y^{(n)}(x)$, then the functions are linearly dependent.

Problem 5 (G. K. Engelis). Let $y^{(1)}(x), \ldots, y^{(m)}(x)$ [$m \leqslant n$] be a system of linearly independent functions on the interval (a, b), where each $y^{(i)}(x)$ has n continuous derivatives. Prove that there exists an equation of the form (5.8) with continuous coefficients having the given functions as particular solutions if and only if the rank of the matrix

$$\left\| \frac{d^k y^{(j)}}{dx^k} \right\| \qquad j = 1, \ldots, m; \, k = 0, \ldots, n - 1$$

equals m at every point of the interval (a, b).[3]

Problem 6. Prove that m functions $y^{(1)}(x), \ldots, y^{(m)}(x)$ continuous on $[a, b]$ are linearly dependent on $[a, b]$ if and only if the *Gram determinant*

$$\det \left\| \int_a^b y^{(i)}(x) y^{(k)}(x) \, dx \right\|$$

vanishes.

37. Reducing the Order of a Homogeneous Linear Equation

Let the functions

$$y^{(1)}(x), \ldots, y^{(m)}(x) \tag{5.13}$$

be m linearly independent solutions of equation (5.8) with continuous coefficients on the interval (a, b). Because of the linear independence, none of the functions (5.13) is identically zero (why?). Let x_0 be any point of (a, b) such that $y^{(1)}(x_0) \neq 0$. Since $y^{(1)}(x)$ is continuous, there is an interval (a_1, b_1) containing x_0 in which

$$|y^{(1)}(x)| > 0.$$

Suppose we change the unknown function in equation (5.8) by setting

$$y(x) = y^{(1)}(x) z(x).$$

---

[3] See the problem at the end of Sec. 35.

Then it is easy to see that $z(x)$ satisfies an equation of the form

$$\frac{d^n z}{dx^n} = a_{n-1}^*(x)\frac{d^{n-1}z}{dx^{n-1}} + a_{n-2}^*(x)\frac{d^{n-2}z}{dx^{n-2}} + \cdots + a_{n-1}^*(x)\frac{dz}{dx} + a_0^*(x)z,$$

where the coefficients $a_i^*(x)$ are continuous on the interval (a_1, b_1). Since $y^{(1)}(x)$ satisfies (5.8), the differential equation for z has the solution

$$z(x) \equiv 1,$$

which in turn implies

$$a_0(x) \equiv 0.$$

Now let

$$\frac{dz}{dx} = y^*.$$

Then the function y^* satisfies the following homogeneous linear equation of order $n - 1$ with continuous coefficients:

$$\frac{d^{n-1}y^*}{dx^{n-1}} = a_{n-1}^*(x)\frac{d^{n-2}y^*}{dx^{n-2}} + a_{n-2}^*(x)\frac{d^{n-3}y^*}{dx^{n-3}} + \cdots + a_1^*(x)y^*. \quad (5.14)$$

Moreover, it is easy to see that each of the functions

$$y_i^*(x) = \frac{d}{dx}\left(\frac{y^{(i+1)}(x)}{y^{(1)}(x)}\right), \qquad i = 1, \ldots, m - 1 \quad (5.15)$$

is a solution of (5.14) on the interval (a_1, b_1). The functions (5.15) are linearly independent on (a_1, b_1). In fact, suppose there exist constants $C_1, \ldots, C_{m-1}$, not all zero, such that the identity

$$\sum_{i=1}^{m-1} C_i \frac{d}{dx}\left(\frac{y^{(i+1)}(x)}{y^{(1)}(x)}\right) \equiv 0$$

holds on (a_1, b_1). Then

$$\sum_{i=1}^{m-1} C_i y^{(i+1)}(x) + C y^{(1)}(x) \equiv 0,$$

where C is a new constant. Therefore the functions (5.13) are linearly dependent on (a_1, b_1). But then some nontrivial linear combination (5.10) vanishes identically on (a_1, b_1), and hence on the interval (a, b) as well, by the uniqueness theorem. In other words, the functions (5.13) are linearly dependent on the entire interval (a, b), contrary to hypothesis. This contradiction shows that the functions (5.15) are linearly independent on (a_1, b_1).

Starting from these $m - 1$ linearly independent solutions of (5.14), we can carry out the same procedure, applied this time to (5.14) instead of (5.8). This gives an equation of order $m - 2$ on some interval (a_2, b_2) contained in (a_1, b_1). Repeating this argument, we finally arrive at a homogeneous linear equation of order $n - m$ on some interval (a_m, b_m).

Problem 1. Prove that with the assumptions made in this section, any interval $(\bar{a}, \bar{b})$ where $a < \bar{a} < \bar{b} < b$ can be decomposed into a finite number of open intervals in each of which the order of the equation can be reduced to $n - m$.

Problem 2. Find the general solution of the equation

$$(2x - 3x^3)y'' + 4y' + 6xy = 0.$$

Hint. One solution is a polynomial in x.

Problem 3. Give a simple method for reducing the number of equations in the system (5.4), starting from a nonzero particular solution.

38. Zeros of Solutions of a Homogeneous Linear Second-Order Equation

Consider the differential equation

$$y'' + a(x)y' + b(x)y = 0, \tag{5.16}$$

where the functions $a(x)$, $a'(x)$ and $b(x)$ are continuous and bounded. In this section, we shall be concerned with the problem (of great practical importance) of how often a solution of (5.16) can vanish. Making the preliminary substitution

$$y(x) = z(x) \exp\left\{-\frac{1}{2}\int_{x_0}^{x} a(\xi)\, d\xi\right\},$$

we reduce (5.16) to the form

$$z'' + B(x)z = 0, \tag{5.17}$$

where the function

$$B = -\frac{a^2}{4} - \frac{a'}{2} + b$$

is itself continuous and bounded. Since $a(x)$ is bounded, the functions $z(x)$ and $y(x)$ vanish at the same time. The following is the basic result of this theory:

THEOREM (*Sturm's theorem*).   *Given two differential equations*

$$z_1''(x) + B_1(x)z_1(x) = 0, \qquad z_2''(x) + B_2(x)z_2(x) = 0,$$

suppose $B_1(x)$ and $B_2(x)$ are continuous on a closed interval $a \leqslant x \leqslant b$ and satisfy the inequality

$$B_2(x) \geqslant B_1(x) \qquad (a \leqslant x \leqslant b). \tag{5.18}$$

Let x_1 and x_2 $(a \leqslant x_1 < x_2 \leqslant b)$ be any two consecutive zeros of a solution $z_1(x) \not\equiv 0$ *of the first equation.*[4] *Then every solution $z_2(x)$ of the second equation has at least one zero between x_1 and x_2, provided that $z_2(x)$ does not vanish for $x = x_1$ and $x = x_2$.*

Proof. Let $z_1(x)$ be a solution of the first differential equation and $z_2(x)$ a solution of the second. Subtracting the second equation multiplied by $z_1(x)$ from the first equation multiplied by $z_2(x)$, we obtain

$$z_1''(x)z_2(x) - z_1(x)z_2''(x) = [B_2(x) - B_1(x)]z_1(x)z_2(x). \qquad (5.19)$$

Noting that

$$z_1''(x)z_2(x) - z_1(x)z_2''(x) = \frac{d}{dx}[z_1'(x)z_2(x) - z_1(x)z_2'(x)],$$

we integrate the identity (5.19) from x_1 to x_2, using the condition

$$z_1(x_1) = z_1(x_2) = 0.$$

This gives

$$z_1'(x_2)z_2(x_2) - z_1'(x_1)z_2(x_1) = \int_{x_1}^{x_2}[B_2(x) - B_1(x)]z_1(x)z_2(x)\, dx. \qquad (5.20)$$

By hypothesis, x_1 and x_2 are consecutive zeros of $z_1(x)$, and hence $z_1(x)$ does not change sign between x_1 and x_2. Since $z_1(x)$ satisfies a linear homogeneous equation, so does $-z_1(x)$, and there is no loss of generality in assuming that $z_1(x)$ is positive between x_1 and x_2. Clearly $z_1'(x_1) \neq 0$, $z_1'(x_2) \neq 0$, since otherwise $z_1(x)$ would vanish identically. This together with the positivity of $z_1(x)$ on (x_1, x_2) implies $z_1'(x_1) > 0$, $z_1'(x_2) < 0$. Suppose there were a solution $z_2(x)$ which is nonvanishing in the open interval (x_1, x_2) and at one (or both) of the end points x_1, x_2, contrary to the conclusion of the theorem. Then we could assume without loss of generality that $z_2(x)$ is nonnegative on $[x_1, x_2]$. But then the left-hand side of (5.20) would be negative while the right-hand side would be nonnegative, since (5.18) implies

$$B_2(x) - B_1(x) \geqslant 0.$$

This contradiction proves the theorem.

COROLLARY 1. *No solution of equation* (5.17) *can vanish in the interval* (a, b) *more than once if $B(x) \leqslant 0$ for all x in (a, b).*

Proof. If a solution $z(x)$ of (5.17) vanished for $x = x_1$ and $x = x_2$ $(a < x_1 < x_2 < b)$, then by Sturm's theorem, *every* solution of the equation $z''(x) = 0$ would have to vanish at least once in the closed interval $[x_1, x_2]$, which is obviously impossible.

---

[4] If $z_1(x)$ has infinitely many zeros in the closed interval $[a, b]$, it is not hard to see that $[a, b]$ contains a point where both $z_1(x)$ and $z_1'(x)$ vanish. But then $z_1(x) \equiv 0$ (cf. Prob. 2, Sec. 36).

COROLLARY 2.   *Let x_1 and x_2 be two consecutive zeros of any solution of equation (5.16). Then every other solution of (5.16) has precisely one zero in the interval (x_1, x_2) if the ratio of the two solutions is not a constant.*

Proof.   Use Sturm's theorem with $B_1(x) \equiv B_2(x)$.

Problem 1.   By comparing the solution of the equation

$$z'' + \left(1 - \frac{n^2 - \frac{1}{4}}{x^2}\right)z = 0$$

(a transformed version of Bessel's equation) with the solutions of the equation $y'' + y = 0$ or $y'' + (1 + \varepsilon^2)y = 0$, show that the distance between consecutive zeros of $z(x) \not\equiv 0$ is less than π for $0 \leqslant n < \frac{1}{2}$, and arbitrarily close to π for sufficiently large x. How are the zeros of $z(x)$ distributed for $n > \frac{1}{2}$?

Problem 2.   Show that consecutive zeros of every solution of the equation

$$y'' + xy = 0$$

become arbitrarily close together as $x \to \infty$.

Problem 3.   In connection with Sturm's theorem, show that if $z_2(x_1) = 0$ and if (5.18) becomes a strict inequality $B_2(x) > B_1(x)$ for at least one point in (x_1, x_2), then the zero of $z_2(x)$ immediately after x_1 lies to the left of x_2.

Problem 4.   Suppose the inequality (5.18) holds, with $z_1(x_1) = z_2(x_1) = 0$ and $z_1'(x_1) \geqslant z_2'(x_1) \geqslant 0$. Prove that $z_1(x) \geqslant z_2(x)$ on the interval from x_1 to the next zero of $z_2(x)$ after x_1, while if there is no such zero, $z_1(x) \geqslant z_2(x)$ for all x in (a, b).

Hint.   First assume that $z_1'(x_1) > z_2'(x_1)$.

Problem 5.   Suppose the function $B(x)$ in equation (5.17) is defined for $0 < a \leqslant x < \infty$ and satisfies the inequality

$$B(x) \geqslant \frac{\frac{1}{4} + \varepsilon}{x^2} \qquad (\varepsilon = \text{const} > 0).$$

Prove that every solution of (5.17) has infinitely many zeros. However, prove that every nontrivial solution has no more than one zero if

$$B(x) \leqslant \frac{1}{4x^2}.$$

Hint.   Make the preliminary change of variables $x = e^t$.

Comment.   For other theorems on oscillations of solutions of (5.17), see Bellman's book.[5]

---

[5] R. Bellman, *Stability Theory of Differential Equations*, McGraw-Hill Book Co., New York (1953).

Problem 6. Consider the equation

$$y'' + B(x, \lambda)y = 0 \qquad (a \leqslant x \leqslant b, \Lambda^* < \lambda < \Lambda^{**}),$$

where the coefficient depends on a parameter λ, and suppose that

a) $B(x, \lambda)$ is continuous in both x and λ;

b) $B(x, \lambda)$ is a nondecreasing function of λ for fixed x;

c) For any $N > 0$, there exists a λ such that

$$B(x, \lambda) < \frac{1}{N} \qquad (a \leqslant x \leqslant b)$$

and a λ such that

$$B(x, \lambda) > N \qquad (a_1 \leqslant x \leqslant b_1, a \leqslant a_1 < b_1 \leqslant b),$$

where a_1 and a_2 are fixed.

Prove that there is a sequence $\Lambda^* < \lambda_1 < \lambda_2 < \cdots < \lambda_n < \cdots$ converging to Λ^{**} such that the equation under consideration has a nontrivial solution for the values $\lambda_1, \lambda_2, \ldots$ and only for these values, and moreover

a) The solution vanishes for $x = a$ and $x = b$;

b) The solution for $\lambda = \lambda_n$ has precisely $n - 1$ zeros between a and b.

39. Systems of Nonhomogeneous Linear First-Order Equations

THEOREM. *Let the vector function $\varphi(x)$ be any particular solution of the nonhomogeneous system (5.2). Then every solution of (5.2) can be represented in the form*

$$y(x) = v(x) + \varphi(x),$$

where $v(x)$ satisfies the homogeneous system (5.4). Conversely, every function $y(x)$ of this form satisfies (5.2).

Proof. The direct assertion follows from

$$\frac{dv}{dx} - A(x)v = \frac{dy}{dx} - A(x)y - \left[\frac{d\varphi}{dx} - A(x)\varphi\right] = f(x) - f(x) = 0,$$

and the converse is proved similarly.

COROLLARY. *Every solution of the nonhomogeneous linear system (5.2) can be represented in the form*

$$y(x) = \varphi(x) + \sum_{k=1}^{n} C_k y^{(k)}(x), \qquad (5.21)$$

where the functions $y^{(1)}(x), \ldots, y^{(n)}(x)$ form a fundamental set of solutions of the corresponding homogeneous system, and $C_1, \ldots, C_n$ are constants (uniquely determined by the given solution). Conversely, the right-hand side of (5.21) satisfies (5.2) for arbitrary $C_1, \ldots, C_n$.

Put somewhat more concisely, *the general solution of a nonhomogeneous linear system is the sum of a particular solution of the system and the general solution of the corresponding homogeneous system.* In other words, the problem of finding the general solution of a nonhomogeneous linear system reduces to finding a particular solution of the system, once we know a fundamental set of solutions of the corresponding homogeneous system. To solve this problem, we now use the method of *variation of constants*, just as in the case of a single linear first-order equation (see Sec. 7).

Thus, given a fundamental set of solutions $y^{(1)}(x), \ldots, y^{(n)}(x)$ of the system (5.4), suppose we try to satisfy the nonhomogeneous system (5.2) by writing

$$y(x) = \sum_{k=1}^{n} C_k(x) y^{(k)}(x), \tag{5.22}$$

where the $C_k(x)$ are now arbitrary differentiable functions instead of constants. Substitution of (5.22) into (5.2) gives

$$\sum_{k=1}^{n} C_k'(x) y^{(k)}(x) + \sum_{k=1}^{n} C_k(x) y^{(k)\prime}(x) - \sum_{k=1}^{n} A(x) C_k(x) y^{(k)}(x)$$

$$= \sum_{k=1}^{n} C_k'(x) y^{(k)}(x) + \sum_{k=1}^{n} C_k(x) [y^{(k)\prime}(x) - A(x) y^{(k)}(x)]$$

$$= \sum_{k=1}^{n} C_k'(x) y^{(k)}(x) = f(x).$$

Writing out the components of this vector equation, we obtain

$$\sum_{k=1}^{n} C_k'(x) y_i^{(k)}(x) = f_i(x), \qquad i = 1, \ldots, n, \tag{5.23}$$

which is a nonhomogeneous linear system in the unknown functions $C_k'(x)$. Moreover, the determinant of the coefficients of these unknowns is nonzero, being the Wronskian of the functions $y_i^{(k)}(x)$. Therefore we can solve (5.23) for the functions

$$C_k'(x) = \psi_k(x), \qquad k = 1, \ldots, n,$$

and then integrate, obtaining

$$C_k(x) = \int \psi_k(x)\, dx + C_k = \Psi_k(x) + C_k, \tag{5.24}$$

where the C_k are arbitrary constants which can be set equal to zero, since we are interested in only one particular solution of the system (5.2). In terms of the functions $\Psi_k(x)$, the required particular solution takes the form

$$y(x) = \sum_{k=1}^{n} \Psi_k(x) y^{(k)}(x).$$

If the C_k are left arbitrary, substitution of (5.24) into (5.22) gives the general solution of (5.2).

Problem 1. Suppose the functions

$$y_i^{(k)}(x, \xi), \qquad i, k = 1, \ldots, n$$

not only satisfy the system (5.4) in x for every fixed k, but also satisfy the following conditions:

$$y_i^{(k)}(\xi, \xi) = \begin{cases} 1 & \text{if } i = k, \\ 0 & \text{if } i \neq k. \end{cases}$$

Prove that the functions

$$y_i(x) = \int_{x_0}^x \sum_{k=1}^n f_k(\xi) y_i^{(k)}(x, \xi)\, d\xi$$

satisfy the system (5.1) and a zero initial condition $y(x_0) = 0$ at any point x_0 where the equations (5.1) are being considered. Use the fundamental matrix $Y(x)$ [see Prob. 1, Sec. 33] to represent this solution in the form

$$y(x) = \int_{x_0}^x Y(x) [\, Y(\xi)]^{-1} f(\xi)\, d\xi.$$

Problem 2. So far we have selected a particular solution from the general solution by specifying an initial condition, i.e., by specifying all the components of the vector function $y(x)$ at the same value of the independent variable x. However, side conditions of a more general nature can be imposed on a system of differential equations. For example, different components of $y(x)$ can be specified at different values of x, or values of linear combinations of the components can be specified at different values of x. Thus suppose the system (5.2) is subject to general linear side conditions which we write concisely as

$$L_k[y] = \alpha_k, \qquad k = 1, \ldots, n,$$

where the α_k are given numbers and the L_k are given combinations of values of the components of $y(x)$ satisfying the linearity condition

$$L_k[C_1 y^{(1)} + C_2 y^{(2)}] = C_1 L_k[y^{(1)}] + C_2 L_k[y^{(2)}].$$

Prove that there are then two possible cases, depending on the coefficient matrix $A(x)$ and on the form of the functionals L_k, i.e., a "basic case" where the problem as posed has precisely one solution for any (continuous) function $f(x)$ and arbitrary values of the α_k, and a "singular case." In the singular case, the problem as a rule has no solution at all for arbitrary $f(x)$ and α_k, but there are infinitely many solutions if a solution exists. Prove that

a) A necessary and sufficient condition for the basic case to occur is that the corresponding homogeneous problem [with $f(x) \equiv 0$ and all $\alpha_k = 0$] have only the solution identically equal to zero;

b) The basic case always occurs for the system (5.2) subject to an initial condition.

Give examples illustrating the singular case.

40. Implications for Nonhomogeneous Linear Equations of Order n

The nonhomogeneous linear equation

$$\frac{d^n y}{dx^n} = a_{n-1}(x)\frac{d^{n-1}y}{dx^{n-1}} + a_{n-2}(x)\frac{d^{n-2}y}{dx^{n-2}} + \cdots + a_1(x)\frac{dy}{dx} + a_0(x)y + f(x)$$
(5.25)

is equivalent to the linear system

$$\frac{dy_0}{dx} = y_1,$$

$$\frac{dy_1}{dx} = y_2,$$

. . . .

$$\frac{dy_{n-2}}{dx} = y_{n-1},$$

$$\frac{dy_{n-1}}{dx} = a_0(x)y_0 + a_1(x)y_1 + a_2(x)y_2 + \cdots + a_{n-1}(x)y_{n-1} + f(x).$$

In this case, the system (5.23) takes the form

$$C_1'(x)y^{(1)} + C_2'(x)y^{(2)} + \cdots + C_n'(x)y^{(n)} = 0,$$

$$C_1'(x)\frac{dy^{(1)}}{dx} + C_2'(x)\frac{dy^{(2)}}{dx} + \cdots + C_n'(x)\frac{dy^{(n)}}{dx} = 0,$$

. . . .

$$C_1'(x)\frac{d^{n-1}y^{(1)}}{dx^{n-1}} + C_2'(x)\frac{d^{n-1}y^{(2)}}{dx^{n-1}} + \cdots + C_n'(x)\frac{d^{n-1}y^{(n)}}{dx} = f(x).$$

Obviously, the only function $y_i(x)$ of interest here is $y_0(x) \equiv y(x)$. Therefore, having determined the functions $C_k(x)$, we need only substitute them into the formula

$$y_0(x) = \sum_{k=1}^{n} C_k(x)y^{(k)}(x).$$

Problem 1. Formulate conditions and prove the assertions of Prob. 2, Sec. 39 as applied to equation (5.25). In particular, analyze the "boundary value problem"

$$y'' + y = f(x) \qquad (a \leqslant x \leqslant b), \qquad y(a) = \alpha_1, \qquad y(b) = \alpha_2,$$

and also the analogous problem for the equation

$$y'' - y = f(x).$$

Problem 2 (Analogue of Prob. 1, Sec. 39). Show that the solution of equation (5.25) with zero initial conditions

$$y(x_0) = y'(x_0) = \cdots = y^{(n-1)}(x_0) = 0$$

is of the form

$$y(x) = \int_{x_0}^{x} f(\xi)G(x, \xi)\,d\xi,$$

where for every fixed ξ the function $G(x, \xi)$ satisfies the corresponding homogeneous equation (5.8) in x and initial conditions

$$y(\xi) = y'(\xi) = \cdots = y^{(n-2)}(\xi) = 0, \qquad y^{(n-1)}(\xi) = 1.$$

Problem 3. Using the result of Prob. 2, prove the following generalization of a theorem due to Chaplygin: For any x_0 in an interval where the coefficients of equation (5.25) are continuous, there are positive numbers $h_0, h_1, \ldots, h_{n-1}$ depending only on x_0 and on the coefficients $a_0(x), a_1(x), \ldots, a_{n-1}(x)$ such that if $y(x)$ is a solution of (5.25) with

$$f(x) \geqslant 0, \quad y(x_0) = y'(x_0) = \cdots = y^{(n-2)}(x_0) = 0, \qquad y^{(n-1)}(x_0) \geqslant 0,$$

then $y^{(k)}(x) \geqslant 0$ $(x_0 \leqslant x \leqslant x_0 + h_k, k = 0, 1, \ldots, n - 1)$, and moreover $y^{(k)}(x) > 0$ $(x_0 < x \leqslant x_0 + h_k)$ if it is also known that $f(x) > 0$ $(x_0 < x \leqslant x_0 + h_k)$. Express the least upper bound of possible values of the numbers h_k in terms of the properties of the function G of Prob. 2. (Can k be set equal to n?). As an example, consider the inequality $y'' + y > 0$, given that $y(0) = 0$, $y'(0) > 0$. On what interval are its solutions guaranteed to be positive? How about derivatives of solutions? Prove that an upper bound for the coefficients $|a_0(x)|, |a_1(x)|, \ldots, |a_{n-1}(x)|$ can be used to deduce a positive lower bound for the numbers h_k. Use Hadamard's lemma to deduce sufficient conditions for the validity of Chaplygin's theorem for nonlinear equations.

6

LINEAR SYSTEMS:
THE CASE OF CONSTANT COEFFICIENTS

---

41. Transformation of a Linear System

In this chapter the unknown functions y_i and the functions $a_{ij}(x)$ and $f_i(x)$ appearing in (5.1) will be allowed to be complex, but we still assume that the independent variable x is real. The real part of a complex function $\varphi(x)$ will be denoted by Re $\varphi(x)$ and the imaginary part by Im $\varphi(x)$. Thus we have

$$\varphi'(x) = \lim_{\Delta x \to 0} \frac{\varphi(x + \Delta x) - \varphi(x)}{\Delta x} = \lim_{\Delta x \to 0} \frac{\text{Re } \varphi(x + \Delta x) - \text{Re } \varphi(x)}{\Delta x}$$

$$+ \, i \lim_{\Delta x \to 0} \frac{\text{Im } \varphi(x + \Delta x) - \text{Im } \varphi(x)}{\Delta x}$$

$$= \frac{d}{dx} \text{Re } \varphi(x) + i \frac{d}{dx} \text{Im } \varphi(x),$$

assuming that the last two derivatives on the right exist. It follows that if C_j and $\varphi_j(x)$ are complex, then

$$\left[\sum_j C_j \varphi_j(x) \right]' = \sum_j C_j \varphi_j'(x),$$

just as for real C_j and $\varphi_j(x)$. Similarly, we can show that derivatives of products of complex functions are given by the usual rules.

From now on we make the additional assumption that the coefficients $a_{ij}(x)$ figuring in (5.1) are constant. Then (5.2) takes the form

$$\frac{dy}{dx} = Ay + f(x), \qquad (6.1)$$

where A is an $n \times n$ coefficient matrix, while $f(x)$ and $\varphi(x)$ are the given and the unknown column vectors. In solving (6.1), our basic idea will be to transform the unknown vector into simpler form by using a suitable linear transformation.

The linear transformation

$$z_i = \sum_{j=1}^{n} k_{ij} y_j, \qquad i = 1, \ldots, n \tag{6.2}$$

can be written in the form

$$z = Ky,$$

where $z = z(x)$ is a new unknown vector, and K is an $n \times n$ matrix characterizing the transformation. Restricting ourselves to nonsingular transformations, for which $\det K \neq 0$, we can solve for y:

$$y = K^{-1}z.$$

Then substituting for y in (6.1), we obtain

$$K^{-1} \frac{dz}{dx} = AK^{-1}z + f(x).$$

It follows that

$$\frac{dz}{dx} = KAK^{-1}z + Kf(x),$$

or finally

$$\frac{dz}{dx} = Bz + g(x), \tag{6.3}$$

where

$$B = KAK^{-1}, \qquad g(x) = Kf(x). \tag{6.4}$$

The system (6.3) has the same form as the system (6.1), except that the coefficient matrix A has been replaced by $B = KAK^{-1}$. Starting from a given matrix A, we naturally try to choose the matrix K in such a way that B takes the simplest possible form. In courses on linear algebra, it is shown that there is always a choice of K such that the matrix B has "Jordan canonical form," in which "Jordan blocks" $\Pi_1, \Pi_2, \ldots, \Pi_k$ $(1 \leqslant k \leqslant n)$ appear along the main diagonal and all the other elements are zero:

$$B = \begin{Vmatrix} \Pi_1 & & & & \\ & \Pi_2 & & & 0 \\ & & \cdot & & \\ & & & \cdot & \\ 0 & & & & \cdot \\ & & & & & \Pi_k \end{Vmatrix}.$$

Every block Π_j is a square matrix of order n_j ($1 \leqslant n_j \leqslant n$) of the form

$$\Pi_j = \begin{Vmatrix} \lambda_j & 1 & & & \\ & \lambda_j & . & & \text{\Large 0} \\ & & . & . & \\ & & & . & 1 \\ \text{\Large 0} & & & . & \lambda_j \end{Vmatrix},$$

where one of the roots λ_j of the *characteristic* (or *secular*) *equation*

$$\det (A - \lambda E) = 0$$

of the matrix A (E is the unit matrix of order n) appears along the main diagonal, the number 1 appears along the diagonal just above the main diagonal, and all the other elements are zero. The order of the matrix Π_j equals the degree of the "elementary divisor "$(\lambda - \lambda_j)^{n_j}$ corresponding to the root λ_j. If a given root has several elementary divisors, each of the corresponding blocks in the Jordan canonical form has the same element along its main diagonal. In the simplest case, where all the elementary divisors are of the first degree (for example, this occurs if all the roots of the characteristic equation are distinct), all the blocks are of order 1, i.e., the matrix B is in purely diagonal form.

In what follows, we shall find it more convenient to use Jordan blocks of a somewhat different form. It is easily verified directly (the reader should do this) that if we set

$$R_j = \begin{Vmatrix} & & & & a_1 \\ \text{\Large 0} & & & a_2 & \\ & & . & & \\ & . & & & \text{\Large 0} \\ a_{n_j} & & & & \end{Vmatrix}$$

where the indicated diagonal contains arbitrary nonzero elements and all the other elements are zero, then

$$R_j \Pi_j R_j^{-1} = \begin{Vmatrix} \lambda_j & & & & \\ a_2 a_1^{-1} & \lambda_j & & & \text{\Large 0} \\ & a_3 a_2^{-1} & . & & \\ & & . & . & \\ & & & . & \\ \text{\Large 0} & & & . & \lambda_j \\ & & & & a_{n_j} a_{n_{j-1}}^{-1} & \lambda_j \end{Vmatrix}.$$

In this way, we can make the numbers under the main diagonal completely arbitrary, thereby obtaining a block of the form

$$
\Pi'_j = \left\|
\begin{array}{cccccc}
\lambda_j & & & & & \\
\gamma_1 & \lambda_j & & & \mathbf{0} & \\
& \gamma_2 & \cdot & & & \\
& & \cdot & \cdot & & \\
& & & \cdot & \cdot & \\
\mathbf{0} & & & & \lambda_j & \\
& & & & \gamma_{n_{j-1}} & \lambda_j
\end{array}
\right\|.
\tag{6.5}
$$

Suppose we subject all the blocks Π_j to transformations of this kind, so that B is transformed according to the formula

$$
B_{\text{new}} = L B_{\text{old}} L^{-1},
$$

where

$$
L = \left\|
\begin{array}{cccc}
R_1 & & & \\
& R_2 & & \mathbf{0} \\
& & \cdot & \\
& & & \cdot \\
\mathbf{0} & & & R_k
\end{array}
\right\|.
$$

Then the matrix B_{new} will be made up of blocks of the form (6.5), and moreover

$$
B_{\text{new}} = L K_{\text{old}} A K_{\text{old}}^{-1} L^{-1} = (L K_{\text{old}}) A (L K_{\text{old}})^{-1} = K_{\text{new}} A K_{\text{new}}^{-1},
$$

i.e., the transformation to blocks of the new form can be accomplished by changing the transformation matrix as indicated.

Suppose such a transformation has already been carried out, so that the matrix B appearing in the system (6.1) consists of Jordan blocks of the form (6.5). Then, going from the matrix notation to the usual scalar notation, we obtain the following canonical form for a system of linear differential equations with constant coefficients:

$$
\frac{dz_1}{dx} = \qquad\qquad\quad + \lambda_1 z_1 \quad\; + g_1(x),
$$

$$
\frac{dz_2}{dx} = \alpha_1 z_1 \qquad\quad + \lambda_1 z_2 \quad\; + g_2(x),
$$

$$
\frac{dz_3}{dx} = \alpha_2 z_2 \qquad\quad + \lambda_1 z_3 \quad\; + g_3(x),
$$

$$
\cdots\cdots\cdots\cdots\cdots\cdots\cdots\cdots\cdots\cdots
$$

$$\frac{dz_{n_1}}{dx} = \alpha_{n_1-1}z_{n_1-1} \quad + \lambda_1 z_{n_1} \quad + g_{n_1}(x),$$

$$\frac{dz_{n_1+1}}{dx} = \qquad\qquad \lambda_2 z_{n_1+1} \quad + g_{n_1+1}(x),$$

$$\frac{dz_{n_1+2}}{dx} = \beta_1 z_{n_1+1} \quad + \lambda_2 z_{n_1+2} \quad + g_{n_1+2}(x),$$

$$\frac{dz_{n_1+3}}{dx} = \beta_2 z_{n_1+2} \quad + \lambda_2 z_{n_1+3} \quad + g_{n_1+3}(x),$$

. . . .

$$\frac{dz_{n_1+n_2}}{dx} = \beta_{n_2-1}z_{n_1+n_2-1} + \lambda_2 z_{n_1+n_2} \quad + g_{n_1+n_2}(x),$$

. . . .

$$\frac{dz_{n-n_k+1}}{dx} = \qquad\qquad \lambda_k z_{n-n_k+1} + g_{n-n_k+1}(x),$$

$$\frac{dz_{n-n_k+2}}{dx} = \omega_1 z_{n-n_k+1} \quad + \lambda_k z_{n-n_k+2} + g_{n-n_k+2}(x),$$

$$\frac{dz_{n-n_k+3}}{dx} = \omega_2 z_{n-n_k+2} \quad + \lambda_k z_{n-n_k+3} + g_{n-n_k+3}(x),$$

. . . .

$$\frac{dz_n}{dx} = \omega_{n_k-1}z_{n-1} \quad + \lambda_k z_n \quad + g_n(x). \tag{6.6}$$

Here the λ_i, α_i, β_i, . . . , ω_i are complex numbers, and the α_i, β_i, . . . , ω_i can be given arbitrary nonzero values provided they are nonzero (in particular, they can all be made arbitrarily small). On the other hand, the λ_i are completely determined by the given system.

After the system has been reduced to canonical form, it can easily be integrated. In fact, the first equation of the "canonical system" involves only one unknown function $z_1(x)$. Determining $z_1(x)$ from this equation and substituting the result into the next equation, we again obtain a linear equation involving only one unknown function $z_2(x)$, and so on.

The integration of linear systems with constant coefficients will be considered in detail, starting from Sec. 45. In Secs. 42–44 we digress to develop the theory of reduction of linear systems to canonical form.

Problem. Prove that Lagrange's formula for finite differences may fail for complex functions of a real variable. However, show that the formula

$$\lambda[f(b) - f(a)] = \mu f'(\xi)$$

holds for a function $f(x)$ differentiable on $a \leqslant x \leqslant b$. Here $a < \xi < b$, λ and μ are real numbers which in general depend on a, b and f, and $\lambda^2 + \mu^2 \neq 0$.

42. Reduction to Canonical Form

THEOREM. *Given a system of linear differential equations*

$$\frac{dy_i}{dx} = \sum_{j=1}^{n} a_{ij} y_j + f_i(x), \qquad i = 1, \ldots, n \qquad (6.7)$$

with constant coefficients a_{ij}, there exists a linear transformation

$$y_i = \sum_{j=1}^{n} c_{ij} z_j$$

with constant coefficients c_{ij} and determinant $|c_{ij}| \neq 0$ which reduces (6.7) *to the canonical form* (6.6), *where the functions $g_i(x)$ in the new system are linear combinations with constant coefficients of the functions $f_i(x)$.*

Proof. The theorem is obviously true for $n = 1$. Assuming that the theorem is true for $n - 1$ equations, we now show that it is also true for n equations, thereby establishing the proof by induction.

We begin by multiplying the ith equation of the system by k_i ($i = 1, \ldots, n$), where the k_i are certain constants to be determined later. We then add the resulting equations, obtaining

$$\frac{d}{dx} \sum_{i=1}^{n} k_i y_i = \sum_{i,j=1}^{n} a_{ij} k_i y_j + \sum_{i=1}^{n} k_i f_i(x).$$

The k_i are now chosen to make the relation

$$\sum_{i,j=1}^{n} a_{ij} k_i y_j = \lambda \sum_{i=1}^{n} k_i y_i \equiv \lambda \sum_{j=1}^{n} k_j y_j$$

an identity in the y_j, where λ is a real or complex number. Obviously, a necessary and sufficient condition for this relation to be an identity is that coefficients of identical y_j be the same in both sides, i.e., that

$$\sum_{i=1}^{n} a_{ij} k_i = \lambda k_j, \qquad j = 1, \ldots, n.$$

Thus the k_i are determined by a system of n homogeneous linear equations in n unknowns. A necessary and sufficient condition for this system to have a nontrivial solution (and this is the only case of interest here) is that the determinant of its coefficient matrix vanish, i.e., that

$$\det\left(\|a_{ij}\| - \lambda E\right) = (-1)^n \det\left(\lambda E - \|a_{ij}\|\right) = 0, \qquad (6.8)$$

where E is the unit matrix of order n. Equation (6.8) is called the *characteristic* (or *secular*) equation, and plays an important role in many problems of mathematics, physics and astronomy. Correspondingly, the matrix $\lambda E - \|a_{ij}\|$ is called the *characteristic matrix* of the system (6.7).

Now let λ_1 be one of the roots of equation (6.8), and let k_{1i} ($i = 1, \ldots, n$) denote any set of numbers, not all zero, satisfying the system of equations

$$\sum_{i=1}^n a_{ij} k_{1i} = \lambda_1 k_{1j} \qquad (j = 1, \ldots, n).$$

Suppose $k_{11} \neq 0$, an assumption which entails no loss of generality since it can always be achieved by suitable renumbering the y_i (the corresponding transformation is nonsingular). Moreover, let

$$z_1 = \sum_{j=1}^n k_{1j} y_j. \qquad (6.9)$$

The function z_1 satisfies the equation

$$\frac{dz_1}{dx} = \lambda_1 z_1 + g_1(x),$$

where

$$g_1(x) = \sum_{i=1}^n k_{1i} f_i(x).$$

We write this equation instead of the first equation of the system (6.7), and then rewrite all the other equations of (6.7), with y_1 replaced by the expression involving $z_1, y_2, \ldots, y_n$ implied by (6.9). Such an expression exists, since $k_{11} \neq 0$ by hypothesis. This gives a new system, which we denote by (6.7′), of the form

$$\frac{dz_1}{dx} = \lambda_1 z_1 \qquad\qquad\qquad\qquad + g_1(x),$$

$$\frac{dy_2}{dx} = a_{21}^* z_1 + a_{22}^* y_2 + a_{23}^* y_3 + \cdots + a_{2n}^* y_n + f_2(x),$$

$$\frac{dy_3}{dx} = a_{31}^* z_1 + a_{32}^* y_2 + a_{33}^* y_3 + \cdots + a_{3n}^* y_n + f_3(x), \qquad (6.7′)$$

$$\cdots\cdots\cdots\cdots\cdots\cdots\cdots\cdots\cdots$$

$$\frac{dy_n}{dx} = a_{n1}^* z_1 + a_{n2}^* y_2 + a_{n3}^* y_3 + \cdots + a_{nn}^* y_n + f_n(x).$$

Next, assuming that our theorem has been proved for $n-1$ equations, we apply it to the system obtained from (6.7′) by omitting the first equation, at the same time regarding $z_1 = z_1(x)$ as a known function just like the $f_i(x)$. Then there exists a nonsingular linear transformation

$$y_i = \sum_{j=2}^{n} k_{ij} y_j^*, \qquad i = 2, \ldots, n,$$

reducing the system (6.7′) to the form

$$\frac{dz_1}{dx} = \lambda_1 z_1 + g_1(x),$$

$$\frac{dy_2^*}{dx} = b_2 z_1 + \lambda_2 y_2^* + \tilde{f}_2(x),$$

$$\frac{dy_3^*}{dx} = b_3 z_1 + \alpha_1 y_2^* + \lambda_2 y_3^* + \tilde{f}_3(x),$$

$$\frac{dy_4^*}{dx} = b_4 z_1 + \alpha_2 y_3^* + \lambda_2 y_4^* + \tilde{f}_4(x),$$

$$\cdots \cdots \cdots \cdots \cdots$$

$$\frac{dy_{n_1+1}^*}{dx} = b_{n_1+1} z_1 + \alpha_{n_1-1} y_{n_1}^* + \lambda_2 y_{n_1+1}^* + \tilde{f}_{n_1+1}(x),$$

$$\frac{dy_{n_1+2}^*}{dx} = b_{n_1+2} z_1 + \lambda_3 y_{n_1+2}^* + \tilde{f}_{n_1+2}(x),$$

$$\frac{dy_{n_1+3}^*}{dx} = b_{n_1+3} z_1 + \beta_1 y_{n_1+2}^* + \lambda_3 y_{n_1+3}^* + \tilde{f}_{n_1+3}(x),$$

$$\frac{dy_{n_1+4}^*}{dx} = b_{n_1+4} z_1 + \beta_2 y_{n_1+3}^* + \lambda_3 y_{n_1+4}^* + \tilde{f}_{n_1+4}(x), \qquad (6.7'')$$

$$\cdots \cdots \cdots \cdots \cdots$$

$$\frac{dy_{n_1+n_2+1}^*}{dx} = b_{n_1+n_2+1} z_1 + \beta_{n_2-1} y_{n_1+n_2}^* + \lambda_3 y_{n_1+n_2+1}^* + \tilde{f}_{n_1+n_2+1}(x),$$

$$\cdots \cdots \cdots \cdots \cdots$$

$$\frac{dy_{n-n_k+1}^*}{dx} = b_{n-n_k+1} z_1 + \lambda_{k+1} y_{n-n_k+1}^* + \tilde{f}_{n-n_k+1}(x),$$

$$\frac{dy_{n-n_k+2}^*}{dx} = b_{n-n_k+2} z_1 + \omega_1 y_{n-n_k+1}^* + \lambda_{k+1} y_{n-n_k+2}^* + \tilde{f}_{n-n_k+2}(x),$$

$$\frac{dy_{n-n_k+3}^*}{dx} = b_{n-n_k+3} z_1 + \omega_2 y_{n-n_k+2}^* + \lambda_{k+1} y_{n-n_k+3}^* + \tilde{f}_{n-n_k+3}(x),$$

$$\cdots \cdots \cdots \cdots \cdots$$

$$\frac{dy_n^*}{dx} = b_n z_1 + \omega_{n_k-1} y_{n-1}^* + \lambda_{k+1} y_n^* + \tilde{f}_n(x).$$

To reduce (6.7″) to canonical form, all that remains is to eliminate some of the b_i. Since every group of equations from the second to the $(n_1 + 1)$st, from the $(n_1 + 2)$nd to the $(n_1 + n_2 + 1)$st, ..., from the $(n_1 - n_k + 1)$st to the nth has the same structure, we shall only show how to eliminate some of the constants $b_2, b_3, \ldots, b_{n_1+1}$. Here two cases must be distinguished, depending on whether or not $\lambda_1 = \lambda_2$.

Case 1. If $\lambda_1 \neq \lambda_2$, we set $z_2 = y_2^* + Cz_1$. Then

$$\frac{dz_2}{dx} = \frac{dy_2^*}{dx} + C\frac{dz_1}{dx} = b_2 z_1 + \lambda_2 y_2^* + C\lambda_1 z_1 + g_2(x)$$

$$= b_2 z_1 + \lambda_2 z_2 - C\lambda_2 z_1 + C\lambda_1 z_1 + g_2(x)$$

$$= \lambda_2 z_2 + [b_2 + C(\lambda_1 - \lambda_2)]z_1 + g_2(x),$$

where $g_2(x)$ is some linear combination of $g_1(x)$ and $\tilde{f}_2(x)$. Let C be such that

$$b_2 + C(\lambda_1 - \lambda_2) = 0.$$

This is possible, since $\lambda_1 \neq \lambda_2$ by hypothesis. Then we obtain

$$\frac{dz_2}{dx} = \lambda_2 z_2 + g_2(x),$$

i.e., b_2 has now been eliminated from the second equation.

Next let $z_3 = y_3^* + C_1 z_1$. Then

$$\frac{dz_3}{dx} = \frac{dy_3^*}{dx} + C_1\frac{dz_1}{dx} = b_3 z_1 + \alpha_1 y_2^* + C_1\lambda_1 z_1 + g_3(x)$$

$$= (b_3 - C\alpha_1 + C_1\lambda_1 - C_1\lambda_2)z_1 + \alpha_1 z_2 + \lambda_2 z_3 + g_3(x),$$

where $g_3(x)$ is some linear combination of $g_1(x)$ and $\tilde{f}_3(x)$. Let C_1 be such that

$$b_3 - C\alpha_1 = C_1(\lambda_2 - \lambda_1),$$

which is again possible, since $\lambda_1 \neq \lambda_2$. As a result, we obtain

$$\frac{dz_3}{dx} = \alpha_1 z_2 + \lambda_2 z_3 + g_3(x),$$

and in the same way, we eliminate the constants b_i from the rest of the equations in the first group.

Case 2. If $\lambda_1 = \lambda_2$, we set

$$y_{n_1+1}^* = z_{n_1+1}, \qquad b_{n_1+1} z_1 + \alpha_{n_1-1} y_{n_1}^* = \alpha_{n_1}^* z_{n_1},$$

where $\alpha_{n_1}^*$ is any nonzero constant. The second of these equations can be solved for $y_{n_1}^*$ and z_{n_1}, since $\alpha_{n_1}^* \neq 0$ and $\alpha_{n_1-1} \neq 0$. Then the

$(n_1 + 1)$st and n_1th equations can be written in the form

$$\frac{dz_{n_1+1}}{dx} = \alpha^*_{n_1} z_{n_1} + \lambda_2 z_{n_1+1} + \tilde{f}_{n_1+1}(x),$$

$$\frac{dz_{n_1}}{dx} = \frac{b_{n_1+1}}{\alpha^*_{n_1}} \frac{dz_1}{dx} + \frac{\alpha_{n_1-1}}{\alpha^*_{n_1}} \frac{dy^*_{n_1}}{dx}$$

$$= \frac{b_{n_1+1}}{\alpha^*_{n_1}} \lambda_1 z_1 + \frac{\alpha_{n_1-1} b_{n_1} z_1}{\alpha^*_{n_1}}$$

$$+ \frac{\alpha_{n_1-1}\alpha_{n_1-2}}{\alpha^*_{n_1}} y^*_{n_1-1} + \frac{\alpha_{n_1-1}}{\alpha^*_{n_1}} \lambda_2 y^*_{n_1} + g_{n_1}(x)$$

$$= \frac{b_{n_1+1}\lambda_1}{\alpha^*_{n_1}} z_1 + \frac{\alpha_{n_1-1} b_{n_1}}{\alpha^*_{n}} z_1 + \frac{\alpha_{n_1-1}\alpha_{n_1-2}}{\alpha^*_{n_1}} y^*_{n_1-1}$$

$$+ \lambda_2 \frac{\alpha^*_{n_1} z_{n_1} - b_{n_1+1} z_1}{\alpha^*_{n_1}} + g_{n_1}(x)$$

$$= \frac{\alpha_{n_1-1} b_{n_1}}{\alpha^*_{n_1}} z_1 + \frac{\alpha_{n_1-1}\alpha_{n_1-2}}{\alpha^*_{n_1}} y^*_{n_1-1} + \lambda_2 z_{n_1} + g_{n_1}(x).$$

If we write

$$\alpha^*_{n_1-1} z_{n_1-1} = \frac{\alpha_{n_1-1} b_{n_1}}{\alpha^*_{n_1}} z_1 + \frac{\alpha_{n_1-1}\alpha_{n_1-2}}{\alpha^*_{n_1}} y^*_{n_1-1},$$

where $\alpha^*_{n_1-1}$ is any nonzero constant, the n_1th equation becomes

$$\frac{dz_{n_1}}{dx} = \alpha^*_{n_1-1} z_{n_1-1} + \lambda_2 z_{n_1} + g_{n_1}(x),$$

and the remaining equations in the first group can be treated similarly. In this way, we get rid of $b_{n_1+1}, b_{n_1}, \ldots, b_4, b_3$. We cannot eliminate b_2 (unless $b_2 = 0$ from the outset), but for $\lambda_1 = \lambda_2$ this is not necessary for reduction to canonical form. However, if $b_2 \neq 0$, the substitution $z_1 = C z^*_1$ can be used to make b_2 arbitrarily small. We remind the reader that $g_i(x)$ always denotes a linear combination of $g_1(x)$ and $\tilde{f}_i(x)$.

Now suppose that in addition to λ_2, there is another λ, say λ_3, which equals λ_1. Then $b_{n_1+3}, b_{n_1+4}, \ldots, b_{n_1+n_2+1}$ can be eliminated by the same method as before. To avoid introducing new notation, we shall assume that

$$b_3 = b_4 = \cdots = b_{n_1+1} = b_{n_1+3} = b_{n_1+4} = \cdots = b_{n_1+n_2+1} = 0$$

already holds in (6.7″). However, b_2 and b_{n_1+2} may be nonzero. If $b_2 = 0$, we can interchange the groups of equations corresponding to λ_2 and λ_3, and then it will be impossible to eliminate b_{n_1+2} in the reduction of the system to canonical form (unless $b_{n_1+2} = 0$ from the outset). On the other hand, if both $b_2 \neq 0$ and $b_{n_1+2} \neq 0$, assuming that $n_1 \geqslant n_2$ (which can always be achieved by interchanging the groups of equations corresponding to λ_2 and λ_3), we set $z_{n_1+2} = y^*_{n_1+2} + C_1 y^*_2$. This gives

$$\frac{dz_{n_1+2}}{dx} = \frac{dy^*_{n_1+2}}{dx} + C_1 \frac{dy^*_2}{dx}$$
$$= b_{n_1+2} z_1 + \lambda_3 y^*_{n_1+2} + C_1 b_2 z_1 + C_1 \lambda_2 y^*_2 + g_{n_1+2}(x)$$
$$= b_{n_1+2} z_1 + \lambda_3 z_{n_1+2} - \lambda_3 C_1 y^*_2 + C_1 b_2 z_1 + C_1 \lambda_2 y^*_2 + g_{n_1+2}(x).$$

Since $b_2 \neq 0$ by hypothesis, we can always choose C_1 such that $C_1 b_2 = -b_{n_1+2}$, obtaining

$$\frac{dz_{n_1+2}}{dx} = \lambda_3 z_{n_1+2} + g_{n_1+2}(x),$$

since $\lambda_2 = \lambda_3$. Moreover, replacing $y^*_{n_1+2}$ by $z_{n_1+2} - C_1 y^*_2$ in the next equation, we find that

$$\frac{dy^*_{n_1+3}}{dx} = -\beta_1 C_1 y^*_2 + \beta_1 z_{n_1+2} + \lambda_3 y^*_{n_1+3} + \tilde{f}_{n_1+3}(x),$$

and the term in y^*_2 can be eliminated from this equation by the substitution $z_{n_1+3} = y^*_{n_1+3} + C_2 y^*_3$. Making repeated transformations of this kind, we eventually reduce the system to canonical form.

Finally we note that all the linear transformations made in reducing the system to canonical form are uniquely invertible, i.e., the linear relations connecting the new variables with the old variables are such that the old variables can be uniquely determined from a knowledge of the new variables. Therefore the transformation carrying the y_i into the z_i is itself linear and invertible, and hence nonsingular. The proof of the theorem is now complete.

Remark. From a practical point of view, the method just described for reducing the system of differential equations (6.7) to canonical form is very formidable. Therefore it is desirable to find methods which more rapidly reveal the structure of the canonical system (i.e., the numbers λ_i and the number of equations corresponding to each λ_i). Such methods will be presented in the next two sections.

43. Invariants of the Characteristic Matrix

THEOREM. *Suppose the nonsingular linear transformation*

$$z_i = \sum_{j=1}^{n} k_{ij} y_j, \qquad i = 1, \ldots, n$$

carries the system

$$\frac{dy_i}{dx} = \sum_{j=1}^{n} a_{ij} y_j, \qquad i = 1, \ldots, n \qquad (6.10)$$

into

$$\frac{dz_i}{dx} = \sum_{j=1}^{n} b_{ij} z_j, \qquad i = 1, \ldots, n. \qquad (6.11)$$

Then

$$\lambda E - B = K(\lambda E - A)K^{-1}, \qquad (6.12)$$

where

$$\|a_{ij}\| = A, \qquad \|b_{ij}\| = B, \qquad \|k_{ij}\| = K$$

and E is the unit matrix of order n. In particular,

$$\det (\lambda E - B) = \det (\lambda E - A),$$

i.e., the determinant of the λ-matrix λE − A is invariant under the transformation.[1] Moreover, the greatest common divisor (g.c.d.) of all minors of order l (l = 1, . . . , n) of the characteristic matrix of the system is invariant under the transformation, in the sense that the g.c.d. of all minors of order l of λE − B is the same as the g.c.d. of all minors of order l of λE − A, to within a constant factor.[2]

Proof. The first assertion is almost obvious: Replacing every z_i in (6.11) by its expression in terms of $y_1, \ldots, y_n$, we obtain

$$\sum_{j=1}^{n} k_{ij} \frac{dy_j}{dx} = \sum_{j=1}^{n} b_{ij} \sum_{s=1}^{n} k_{js} y_s,$$

and hence, after substituting from (6.10),

$$\sum_{j=1}^{n} k_{ij} \sum_{s=1}^{n} a_{js} y_s = \sum_{j,s=1}^{n} k_{ij} a_{js} y_s = \sum_{j,s=1}^{n} b_{ij} k_{js} y_s.$$

Since this formula must be an identity in the y_s, we have

$$\sum_{j=1}^{n} k_{ij} a_{js} = \sum_{j=1}^{n} b_{ij} k_{js}$$

for every i and s, and hence, in matrix notation,

$$KA = BK \qquad \text{or} \qquad B = KAK^{-1},$$

which is equivalent to (6.12), since $E = KEK^{-1}$. The invariance of the determinant follows from (6.12) and the familiar fact that det $MN =$ det M det N (where M and N are arbitrary $n \times n$ matrices).

---

[1] By a *λ-matrix* is meant a matrix whose elements are polynomials in λ.
[2] In calculating the g.c.d., think of the minors as polynomials.

The proof of the invariance of the g.c.d. is a bit harder. Let $a_l(\lambda)$ be the g.c.d. of all minors of order l of $\lambda E - A$, and let $b_l(\lambda)$ be the g.c.d. of all minors of order l of $\lambda E - B$. Every minor of order l of the product $K(\lambda E - A)$ is a sum of products of minors of order l of K with minors of order l of $\lambda E - A$.[3] Therefore every common divisor of all the minors of order l of $\lambda E - A$ divides every minor of order l of $K(\lambda E - A)$. Similarly, every common divisor of all the minors of order l of $\lambda E - A$ divides every minor of order l of $K(\lambda E - A)K^{-1} = \lambda E - B$, and hence divides $b_l(\lambda)$. But then $a_l(\lambda)$ divides $b_l(\lambda)$. Reversing the roles of A and B, we see that $b_l(\lambda)$ divides $a_l(\lambda)$. It follows that $a_l(\lambda)$ and $b_l(\lambda)$ agree to within a constant factor, and the theorem is proved.

Problem. Prove the *Hamilton-Cayley theorem*: Every matrix satisfies its own characteristic equation, i.e., $\lambda E - A$ reduces to the zero matrix if λ is replaced by the matrix A.

44. Elementary Divisors

LEMMA.[4]   *Suppose a square matrix P of order n contains a rectangular submatrix Q of height a and width b (i.e., with a rows and b columns) whose elements are all zero. Then $\det P = 0$ if $a + b > n$.*

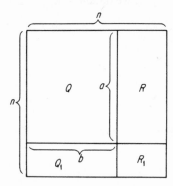

FIGURE 25

Proof. We can always move Q into the upper left-hand corner of P without changing the absolute value of $\det P$. Then P has the structure shown schematically in Figure 25. By Laplace's theorem,[5] the determinant of P is a sum of products of minors of P of order b made up of elements in the rectangular matrices Q and Q_1 with corresponding signed minors made up of elements in the rectangular matrices R and R_1. But since $b > n - a$, each of these minors of order b contains at least one row consisting entirely of zeros. Therefore $\det P = 0$, as asserted.

[3] For the explicit formula, see e.g., V. I. Smirnov, *Linear Algebra and Group Theory* (translated by R. A. Silverman), McGraw-Hill Book Co., New York (1961), p. 30.

[4] Suggested by S. L. Sobolev.

[5] See e.g., V. I. Smirnov, *op. cit.*, p. 17.

Now consider the matrix

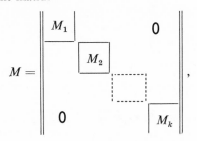

$$M = \begin{Vmatrix} M_1 & & & & 0 \\ & M_2 & & & \\ & & \ddots & & \\ 0 & & & & M_k \end{Vmatrix},$$

where

$$M_s = \begin{Vmatrix} \lambda - \lambda_s & & & & & \\ \varepsilon_{s1} & \lambda - \lambda_s & & & 0 & \\ & \varepsilon_{s2} & \lambda - \lambda_s & & & \\ & & \cdot & \cdot & & \\ & & & \cdot & \cdot & \\ & & & & \cdot & \\ & & & \lambda \cdot & - \lambda_s & \\ 0 & & & \varepsilon_{sn_s-2} & \lambda - \lambda_s & \\ & & & & \varepsilon_{sn_s-1} & \lambda - \lambda_s \end{Vmatrix}.$$

The matrix M_s $(s = 1, \ldots, k)$ is of order n_s, and the numbers $\varepsilon_{s1}, \varepsilon_{s2}, \ldots,$ ε_{sn_s-1} are all nonzero. Some of the numbers $\lambda_1, \lambda_2, \ldots, \lambda_k$ may coincide, and all elements except the ε_{si} and $\lambda - \lambda_s$ are zero. The matrix M will be recognized as the characteristic matrix associated with the canonical form (6.6) of the system of differential equations (6.7), except for a slight change in the notation used for the off-diagonal elements.

Obviously the determinant of M equals the product of the determinants of the submatrices $M_1, M_2, \ldots, M_s$, and hence

$$\det M = \det M_1 \det M_2 \cdots \det M_s = (\lambda - \lambda_1)^{n_1}(\lambda - \lambda_2)^{n_2} \cdots (\lambda - \lambda_k)^{n_k}.$$

Therefore the results of the preceding two sections imply

THEOREM 1.   *The numbers $\lambda_1, \lambda_2, \ldots, \lambda_k$ appearing in the equations (6.6) are the roots of the characteristic equation of the system (6.7).*

Next we find the g.c.d. of all minors of order l of the λ-matrix M:

THEOREM 2.   *Let $D_l(\lambda)$ be the g.c.d. of all minors of order l ($l = 1, \ldots, n$) of the matrix M, and let m_i denote the sum of the orders of all the submatrices M_s with diagonal element $\lambda - \lambda^{(i)}$. Moreover, let $m_i^{(1)}$ be the largest order of all submatrices with diagonal elements $\lambda - \lambda^{(i)}$,* *let $m_i^{(2)}$ be the second largest order of all such submatrices, and so on. Then, to within a constant factor,*

$$D_l(\lambda) = \prod_{i=1}^{m} (\lambda - \lambda^{(i)})^{m_i - m_i^{(1)} - \cdots - m_i^{(n-l)}},$$

where the distinct values of $\lambda_1, \lambda_2, \ldots, \lambda_k$ are denoted by

$$\lambda^{(1)}, \lambda^{(2)}, \ldots, \lambda^{(m)} \qquad (m \leqslant k).$$

Proof. The fact that the expression for $D_l(\lambda)$ is valid for $l = n$ has just been proved. Thus let $l = n - 1$. Since a common divisor of all minors of order l of the matrix M must also be a divisor of det M, we have

$$D_l(\lambda) = (\lambda - \lambda^{(1)})^{p_1}(\lambda - \lambda^{(2)})^{p_2} \ldots (\lambda - \lambda^{(m)})^{p_m},$$

where the p_i are appropriate nonnegative integers ($p_i \leqslant m_i$). Suppose that from the matrix M we delete a row and column which intersect outside all the submatrices M_s, as shown schematically in Figure 26. Then the remaining matrix M' of order $n - 1$ contains a rectangular submatrix Q, the sum of whose height and width equals n (Q is the shaded area in the figure, for the case where the deleted row intersects M_1). Therefore det $M' = 0$, by the lemma, and hence to find the g.c.d. $D_{n-1}(\lambda)$ of all minors of order $n - 1$ of the matrix M, we need only consider matrices obtained from M by deleting rows and columns intersecting inside one of the matrices M_s, say M_{s_1}. Clearly the determinant of such a matrix M' equals the determinant of the matrix M'_{s_1}, obtained from M_{s_1} by deleting one row and one column, multiplied by the determinants of all the other matrices M_s. Moreover, in finding the g.c.d. of all minors of order $n - 1$ of M, the minors of key importance are those containing the smallest powers of the factors $\lambda - \lambda^{(1)}, \ldots, \lambda - \lambda^{(m)}$. But the smallest power of $\lambda - \lambda_{s_1}$ in the determinant of M'_{s_1} is obtained by deleting the first row and the last column, since then the determinant is just the constant

$$\varepsilon_{s_1 1}\varepsilon_{s_2 2} \ldots \varepsilon_{s_1 n_{s_1} - 1},$$

which is nonzero, since all the $\varepsilon_{s_1 i}$ are nonzero. In this case, $\lambda - \lambda_{s_1}$ appears in det M' to a power n_{s_1} less than in det M. Therefore the smallest power of $\lambda - \lambda^{(i)}$ which can appear in all the minors of order $n - 1$ of the matrix M equals

$$(\lambda - \lambda^{(i)})^{m_i - m_i^{(1)}},$$

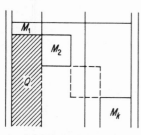

FIGURE 26

and hence

$$D_{n-1}(\lambda) = \prod_{i=1}^{m} (\lambda - \lambda^{(i)})^{m_i - m_i^{(1)}}.$$

This completes the proof for $l = m - 1$. Similarly

$$D_{n-2}(\lambda) = \prod_{i=1}^{m} (\lambda - \lambda^{(i)})^{m_i - m_i^{(1)} - m_i^{(2)}},$$

and more generally we obtain the formula given in the statement of the theorem.

DEFINITION.   *The factors*

$$(\lambda - \lambda^{(i)})^{m_i^{(j)}}$$

are called elementary divisors.

According to Sec. 43, the g.c.d. of all minors of order l of the characteristic matrix $\lambda E - A$ corresponding to the system of differential equations (6.7) is the same (to within a constant factor) as the g.c.d. of all minors of the characteristic matrix M corresponding to the system after reduction to canonical form. Therefore $\lambda E - A$ and M have the same elementary divisors. In particular, if $\lambda E - A$ has several identical elementary divisors, so does M. Thus the structure of the canonical form of the matrix $\lambda E - A$ can be inferred from a knowledge of its elementary divisors and their multiplicities (the number of times each occurs). Only the off-diagonal terms in the matrices M_s remain undetermined, and as we saw in Sec. 42, they can be given arbitrary nonzero values.

45. Determination of a Fundamental Set of Solutions of a Homogeneous System

LEMMA.   *The m vector functions $y^{(1)}(x), \ldots, y^{(m)}(x)$ are linearly independent if and only if the vector functions $z^{(1)}(x) = K y^{(1)}(x), \ldots, z^{(m)}(x) = K y^{(m)}(x)$ obtained from $y^{(1)}(x), \ldots, y^{(m)}(x)$ by applying a nonsingular linear transformation K are also linearly independent.*

Proof.   Suppose there are constants $C_1, \ldots, C_m$ not all zero such that

$$\sum_{i=1}^{m} C_i z^{(i)}(x) \equiv 0.$$

Then

$$\sum_{i=1}^{m} C_i K y^{(i)}(x) = K \sum_{i=1}^{m} C_i y^{(i)}(x) \equiv 0,$$

and hence premultiplying by K^{-1} we obtain

$$\sum_{i=1}^{m} C_i y^{(i)}(x) \equiv 0,$$

which contradicts the assumed linear independence of the functions
$y^{(i)}(x)$. The converse is proved by interchanging the roles of $y^{(i)}(x)$
and $z^{(i)}(x)$ and those of K and K^{-1}.

Now consider the homogeneous system

$$\frac{dy}{dx} = Ay,$$

or in scalar form

$$\frac{dy_i}{dx} = \sum_{j=1}^{n} a_{ij} y_j, \qquad i = 1, \dots, n. \tag{6.13}$$

Then, as we have seen above, the elementary divisor $(\lambda - \lambda_s)^{p_s}$ of the matrix
$\lambda E - A$ corresponds to the following group of homogeneous equations in
the canonical system (6.6), involving nonzero constants $\varepsilon_1, \varepsilon_2, \dots, \varepsilon_{p_s-1}$:

$$\frac{dz_{k+1}}{dx} = \lambda_s z_{k+1},$$

$$\frac{dz_{k+2}}{dx} = \varepsilon_1 z_{k+1} + \lambda_s z_{k+2},$$

$$\frac{dz_{k+3}}{dx} = \varepsilon_2 z_{k+2} + \lambda_s z_{k+3}, \tag{6.14}$$

$$\cdots \cdots \cdots \cdots \cdots \cdots$$

$$\frac{dz_{k+p_s-1}}{dx} = \varepsilon_{p_s-2} z_{k+p_s-2} + \lambda_s z_{k+p_s-1},$$

$$\frac{dz_{k+p_s}}{dx} = \varepsilon_{p_s-1} z_{k+p_s-1} + \lambda_s z_{k+p_s}.$$

Suppose we introduce new unknown functions by writing

$$z_{k+l} = z_{k+l}^{*} e^{\lambda_s x},$$

where in general λ_s is complex, with real part λ_s' and imaginary part λ_s''.[6]

---

[6] Thus $e^{\lambda_s x}$ means

$$e^{\lambda_s' x}(\cos \lambda_s'' x + i \sin \lambda_s'' x)$$

(Euler's formula), and

$$\frac{d}{dx} e^{\lambda_s x} = \lambda_s' e^{\lambda_s' x}(\cos \lambda_s'' x + i \sin \lambda_s'' x) + e^{\lambda_s' x}(-\lambda_s'' \sin \lambda_s'' x + i\lambda_s'' \cos \lambda_s'' x)$$

$$= (\lambda_s' + i\lambda_s'') e^{\lambda_s' x}(\cos \lambda_s'' x + i \sin \lambda_s'' x) = \lambda_s e^{\lambda_s x},$$

just as in the case of real λ_s.

After making this substitution, we find that

$$\frac{dz_{k+1}^*}{dx} = 0,$$

$$\frac{dz_{k+2}^*}{dx} = \varepsilon_1 z_{k+1}^*,$$

$$\frac{dz_{k+3}^*}{dx} = \varepsilon_2 z_{k+2}^*,$$

$$\cdot \quad \cdot \quad \cdot \quad \cdot \quad \cdot \quad \cdot \quad \cdot \quad \cdot \quad \cdot$$

$$\frac{dz_{k+p_s-1}^*}{dx} = \varepsilon_{p_s-2} z_{k+p_s-2}^*,$$

$$\frac{dz_{k+p_s}^*}{dx} = \varepsilon_{p_s-1} z_{k+p_s-1}^*.$$

The solution of these equations is

$$z_{k+1}^* = C_1 = C_0^{(1)}.$$

Substituting z_{k+1}^* into the second equation and integrating, we obtain

$$z_{k+2}^* = C_1 \varepsilon_1 x + C_2 = C_1^{(2)} x + C_0^{(2)}.$$

Then substitution of z_{k+2}^* into the third equation gives

$$z_{k+3}^* = \frac{C_1 \varepsilon_1 \varepsilon_2}{2} x^2 + C_2 \varepsilon_2 x + C_3 = C_2^{(3)} x^2 + C_1^{(3)} x + C_0^{(3)},$$

and so on, up to

$$z_{k+p_s}^* = C_{p_s-1}^{(p_s)} x^{p_s-1} + C_{p_s-2}^{(p_s)} x^{p_s-2} + \cdots + C_1^{(p_s)} x + C_0^{(p_s)}.$$

Here the C's with various indices always denote real or complex constants. Finally, transforming back to the variables $z_{k+1}, z_{k+2}, \ldots, z_{k+p_s}$, we have

$$z_{k+1} = e^{\lambda_s x} C_0^{(1)}$$

$$z_{k+2} = e^{\lambda_s x}(C_1^{(2)} x + C_0^{(2)}),$$

$$z_{k+3} = e^{\lambda_s x}(C_2^{(3)} x^2 + C_1^{(3)} x + C_0^{(3)}), \qquad\qquad (6.15)$$

$$\cdot \quad \cdot \quad \cdot \quad \cdot \quad \cdot \quad \cdot \quad \cdot \quad \cdot \quad \cdot \quad \cdot \quad \cdot \quad \cdot$$

$$z_{k+p_s} = e^{\lambda_s x}(C_{p_s-1}^{(p_s)} x^{p_s-1} + C_{p_s-2}^{(p_s)} x^{p_s-2} + \cdots + C_0^{(p_s)}).$$

The equations (6.15) give the general solution of the system of differential equations (6.14) and moreover, according to Theorem 4, Sec. 33 the system has p_s linearly independent solutions of the form (6.15). We can satisfy the entire homogeneous canonical system by setting all the z_i equal to zero except those appearing in (6.14) and (6.15). Since the y_i are linear combinations of the z_i ($i = 1, \ldots, n$), it follows from the lemma on p. 139 that the system (6.11) has p_s linearly independent solutions of the form

$$y_i = e^{\lambda_s x} \sum_{j=0}^{p_s-1} C_j^{*(i)} x^j, \qquad i = 1, \ldots, n,$$

corresponding to each elementary divisor $(\lambda - \lambda_s)^{p_s}$ of the matrix $\lambda E - A$. It is easy to see that we can assume that

$$C_j^{*(i)} = 0 \quad \text{for} \quad j > 0 \quad \text{but} \quad \sum_{i=1}^{n} |C_0^{*(i)}| > 0$$

in the first of these solutions, corresponding to $z_{k+2} \equiv \cdots \equiv z_{k+p_s} \equiv 0$, that

$$C_j^{*(i)} = 0 \quad \text{for} \quad j > 1 \quad \text{but} \quad \sum_{i=1}^{n} |C_1^{*(i)}| > 0$$

in the second of these solutions, corresponding to $z_{k+3} \equiv \cdots \equiv z_{k+p_s} \equiv 0$, and so on.

If the matrix $\lambda E - A$ has several elementary divisors, equal to various powers of $\lambda - \lambda^{(1)}$, say

$$(\lambda - \lambda^{(1)})^{\mu_1}, (\lambda - \lambda^{(1)})^{\mu_2}, \ldots, (\lambda - \lambda^{(1)})^{\mu_k},$$

then the system (6.11) has $\mu_1 + \cdots + \mu_k$ linearly independent solutions of the form

$$y_i = e^{\lambda^{(1)} x} \sum_{j=0}^{M-1} C_j^{(i)} x^j, \qquad i = 1, \ldots, n,$$

where $M = \max(\mu_1, \ldots, \mu_k)$, which can be found by substituting these expressions into (6.11), dividing by $e^{\lambda^{(1)} x}$ and then comparing coefficients of identical powers of x.[7] These solutions are linearly independent of those found in the same way for the elementary divisors involving powers of other $\lambda - \lambda^{(p)}$, because such elementary divisors are associated with other groups of equations in the canonical system.

Remark. Suppose the system (6.11) has real coefficients a_{ij}, and let $y(x)$ be a complex solution of (6.11). Then Re $y(x)$ and Im $y(x)$ are also solutions of (6.11). Moreover, given n linearly independent complex solutions $y^{(k)}(x)$, $k = 1, \ldots, n$ of (6.11), there must be n linearly independent functions among the $2n$ functions Re $y^{(k)}(x)$, Im $y^{(k)}(x)$, $k = 1, \ldots, n$ (why?). Therefore a homogeneous system with real coefficients always has a fundamental set of *real* solutions.

---

[7] In other words, we use the method of undetermined coefficients to find the $C_j^{(i)}$.

Problem 1. Given a square matrix A of order n, we define

$$e^A = A + \frac{1}{1!}A + \frac{1}{2!}A^2 + \cdots + \frac{1}{n!}A^n + \cdots.$$

Prove that this series converges and that the matrix e^A has the following properties:

$$e^{x_1 A}e^{x_2 A} = e^{(x_1 + x_2)A}, \qquad e^0 = E, \qquad e^{-A} = (e^A)^{-1}.$$

Show that the matrix e^{xA} is differentiable with respect to x and $(e^{xA})' = Ae^{xA}$. Prove that if $A = K^{-1}BK$, then $e^A = K^{-1}e^B K$. Find an expression for e^{xA} in powers of a matrix with (ordinary) Jordan canonical form. Prove that if the real parts of the characteristic equation of the matrix A are less than some number $-\alpha < 0$, then

$$\|e^{-xA}\| \leqslant Me^{-\alpha x}$$

for $x > 0$ (see Prob. 1, Sec. 32), where the constant M depends on the choice of the matrix A.

Problem 2. Prove that the general solution of the system (6.11) is of the form

$$y = e^{xA}C,$$

where C is an arbitrary constant vector. Prove that the solution of (6.11) satisfying the initial condition $y(x_0) = y^0$ is

$$y = e^{(x - x_0)A}\, y^0.$$

46. Implications for Homogeneous Equations of Order n

We now consider the system (5.9), which is equivalent to the single equation (5.8), assuming that all the coefficients $a_0(x), \ldots, a_{n-1}(x)$ are constants $a_0, \ldots, a_{n-1}$. Then the characteristic matrix of (5.9) is

$$\begin{Vmatrix} \lambda & -1 & & & & & \\ & \lambda & -1 & & & & \\ & & \lambda & -1 & & \mathbf{0} & \\ & & & \cdot & \cdot & & \\ & & & & \cdot & \cdot & \\ & & & & & \cdot & \cdot \\ & \mathbf{0} & & & & \lambda & -1 \\ -a_0 & -a_1 & -a_2 & -a_3 & \cdots & -a_{n-2} & \lambda - a_{n-1} \end{Vmatrix}. \qquad (6.16)$$

It is easy to see that this matrix has the determinant

$$M(\lambda) \equiv \lambda^n - a_{n-1}\lambda^{n-1} - a_{n-2}\lambda^{n-2} - \cdots - a_1\lambda - a_0. \qquad (6.17)$$

By deleting the first column and the last row of (6.16), we obtain a matrix whose determinant equals $+1$ or -1. Therefore, according to Theorem 2, p. 137, *if the polynomial $M(\lambda)$ has a root λ_s of multiplicity p_s, the matrix* (6.16) *has the elementary divisor $(\lambda - \lambda_s)^{p_s}$ and no other elementary divisors involving powers of $\lambda - \lambda_s$.*

It now follows that a root λ_s of multiplicity p_s is associated with p_s linearly independent solutions of the form[8]

$$(C_0 + C_1 x + C_2 x^2 + \cdots + C_{p_s-1} x^{p_s-1})e^{\lambda_s x}, \qquad (6.18)$$

which can clearly be chosen to be

$$e^{\lambda_s x}, \, x e^{\lambda_s x}, \, x^2 e^{\lambda_s x}, \, \ldots, \, x^{p_s-1} e^{\lambda_s x}.$$

In fact, if these functions were linearly dependent, then some expression of the form (6.18) with at least one nonzero coefficient C_i would vanish identically. But this is impossible since $e^{\lambda_s x}$ never vanishes and a polynomial cannot be identically zero unless all its coefficients vanish.

Remark. Suppose the coefficients a_i are all real, and let $(\lambda - \lambda_s)^{p_s}$ be an elementary divisor of the matrix (6.16) where λ_s is complex. Then (6.16) also has the complex conjugate elementary divisor $(\lambda - \bar\lambda_s)^{p_s}$. Moreover, if

$$y(x) = x^k e^{\lambda_s x} = x^k e^{\alpha_s x}(\cos \beta_s x + i \sin \beta_s x)$$

$(\lambda_s = \alpha_s + i\beta_s)$ is a solution of equation (5.8) with constant coefficients, then so is

$$\bar y(x) = x^k e^{\bar\lambda_s x} = x^k e^{\alpha_s x}(\cos \beta_s x - i \sin \beta_s x).$$

Therefore the real functions

$$\frac{y(x) + \bar y(x)}{2} = x^k e^{\bar\alpha_s x} \cos \beta_s x,$$

$$\frac{y(x) - \bar y(x)}{2i} = x^k e^{\alpha_s x} \sin \beta_s x$$

both satisfy (5.8), and in this way we obtain n linearly independent real solutions of (5.8) [why?].

Problem 1. Show that if the characteristic matrix of a system of n first-order linear equations with constant coefficients has the property in italics above, then there is a nonsingular linear transformation reducing the system to the form (5.9) equivalent to a single equation of order n.[9]

---

[8] These solutions are also linearly independent of the solutions associated with other roots of $M(\lambda)$, since the latter solutions correspond to other groups of equations in the canonical form of the system (5.9).

[9] In other words, we now have a necessary and sufficient condition for a system of n first-order linear equations with constant coefficients to be equivalent (in the indicated sense) to a single linear equation of order n with constant coefficients.

Problem 2. Find the general solution of Euler's equation

$$x^n \frac{d^n y}{dx^n} + a_{n-1} x^{n-1} \frac{d^{n-1} y}{dx^{n-1}} + \cdots + a_1 x \frac{dy}{dx} + a_0 y = 0,$$

where all the a_i are constant. (Introduce a new independent variable t by setting $x = \pm e^t$.) Note that we can also handle the case where every power of x is replaced by the corresponding power of $ax + b$.

Problem 3. Find all solutions of the equation

$$y'(x) = ay(x) + by(c - x)$$

which exist for $-\infty < x < \infty$ (a, b and c are constants).

Problem 4. Derive all the basic properties of the functions $\sin x$ and $\cos x$ by regarding them as the solutions of the equation $y'' + y = 0$ satisfying the initial conditions $y|_{x=0} = 0$, $y'|_{x=0} = 1$ and $y|_{x=0} = 1$, $y'|_{x=0} = 0$, respectively.

Problem 5. After writing

$$L[f] = \frac{d^n f}{dx^n} - a_{n-1} \frac{d^{n-1} f}{dx^{n-1}} - \cdots - a_1 \frac{df}{dx} - a_0 f,$$

deduce the formula

$$L[e^{\lambda x} f] = e^{\lambda x} \left\{ M(\lambda) f + \frac{M'(\lambda)}{1!} \frac{df}{dx} + \cdots + \frac{M^{(n)}(\lambda)}{n!} \frac{d^n f}{dx^n} \right\}.$$

Use this result to obtain the general solution of equation (5.8) with constant coefficients.

47. Determination of a Particular Solution of a Nonhomogeneous System

In studying nonhomogeneous systems, we confine ourselves to the case where the functions f_i appearing in (6.7) are of the form

$$f_i(x) = \sum_{k=1}^{m} C_i^{(k)} e^{\alpha_k x} x^{\beta_k},$$

with real or complex α_k, $C_i^{(k)}$ and nonnegative integral β_k. Obviously, it is sufficient to study the case $m = 1$, since the particular solution in the general case will be a sum of solutions corresponding to this special case. Thus we set

$$f_i(x) = C_i e^{\alpha x} x^{\beta},$$

and then write the group of equations in the system (6.6) corresponding to a given elementary divisor $(\lambda - \lambda_s)^{p_s}$ of the characteristic matrix $\lambda E - A$:

$$\frac{dz_{k+1}}{dx} = \lambda_s z_{k+1} \qquad\qquad\qquad + C^*_{k+1} x^\beta e^{\alpha x},$$

$$\frac{dz_{k+2}}{dx} = \varepsilon_1 z_{k+1} + \lambda_s z_{k+2} \qquad\quad + C^*_{k+2} x^\beta e^{\alpha x},$$

$$\frac{dz_{k+3}}{dx} = \varepsilon_2 z_{k+2} + \lambda_s z_{k+3} \qquad\quad + C^*_{k+3} x^\beta e^{\alpha x},$$

$$\cdots \cdots \cdots \cdots \cdots \cdots \cdots \cdots \cdots \cdots$$

$$\frac{dz_{k+p_s}}{dx} = \varepsilon_{p_s-1} z_{k+p_s-1} + \lambda_s z_{k+p_s} + C^*_{k+p_s} x^\beta e^{\alpha x}.$$

Here the C^*_i are certain new constants. Using the relations

$$z_i = z^*_i e^{\lambda_s x}, \qquad i = k+1, \ldots, k+p_s$$

to introduce new unknown functions z^*_i, we find that

$$\frac{dz^*_{k+1}}{dx} = \qquad\qquad\qquad C^*_{k+1} x^\beta e^{(\alpha-\lambda_s)x},$$

$$\frac{dz^*_{k+2}}{dx} = \varepsilon_1 z^*_{k+1} \qquad\quad + C^*_{k+2} x^\beta e^{(\alpha-\lambda_s)x},$$

$$\frac{dz^*_{k+3}}{dx} = \varepsilon_2 z^*_{k+2} \qquad\quad + C^*_{k+3} x^\beta e^{(\alpha-\lambda_s)x}, \qquad (6.19)$$

$$\cdots \cdots \cdots \cdots \cdots \cdots \cdots \cdots \cdots$$

$$\frac{dz^*_{k+p_s}}{dx} = \varepsilon_{p_s-1} z^*_{k+p_s-1} + C^*_{k+p_s} x^\beta e^{(\alpha-\lambda_s)x}.$$

In integrating this system, two cases should be distinguished, depending on whether or not λ_s equals α.

Case 1. If $\lambda_s \neq \alpha$ we can integrate the equations (6.19) step by step, obtaining

$$z^*_i = M^{(\beta)}_i(x) e^{(\alpha-\lambda_s)x}, \qquad i = k+1, \ldots, k+p_s,$$

where each $M^{(\beta)}_i(x)$ is a polynomial in x of degree no higher than β.[10] It follows that

$$z_i = M^{(\beta)}_i(x) e^{\alpha x}, \qquad i = k+1, \ldots, k+p_s.$$

---

[10] According to complex variable theory, the formulas obtained by integrating $x^\beta e^{(\alpha-\lambda_s)x}$ for real $\alpha - \lambda_s$ remain valid if $\alpha - \lambda_s$ is complex. This can also be verified directly.

If none of the λ_s equals α, then all the z_i $(i = 1, \ldots, n)$ are of the form

$$z_i = M_i^{(\beta)}(x)e^{\alpha x},$$

and hence the particular solution y_i is also of the form

$$y_i = M_i^{*(\beta)}(x)e^{\alpha x}. \qquad (6.20)$$

The coefficients $M_i^{*(\beta)}(x)$ can be found by substituting (6.20) into (6.7), dividing the resulting equations by $e^{\alpha x}$, and then comparing coefficients of identical powers of x.

Case 2. If $\lambda_s = \alpha$ the system (6.19) becomes

$$\frac{dz_{k+1}^*}{dx} = \qquad\qquad\qquad C_{k+1}^* x^{\beta},$$

$$\frac{dz_{k+2}^*}{dx} = \varepsilon_1 z_{k+1}^* \qquad + C_{k+2}^* x^{\beta},$$

$$\frac{dz_{k+3}^*}{dx} = \varepsilon_2 z_{k+2}^* \qquad + C_{k+3}^* x^{\beta},$$

$$\cdots\cdots\cdots\cdots\cdots\cdots$$

$$\frac{dz_{k+p_s}^*}{dx} = \varepsilon_{p_s-1} z_{k+p_s-1}^* + C_{k+p_s}^* x^{\beta}.$$

Integrating these equations step by step, we obtain a particular solution of the form

$$z_{k+i}^*(x) = M_{k+i}^{(i)}(x)x^{\beta}, \qquad i = 1, \ldots, p_s,$$

where each $M_{k+1}^{(i)}$ is a polynomial in x of degree no higher than i. It follows that

$$z_{k+i}(x) = M_{k+i}^{(i)}x^{\beta}e^{\alpha x}, \qquad i = 1, \ldots, p_s.$$

Therefore the system (6.7) has a particular solution of the form[11]

$$y_i(x) = M_i^{*(\beta+p)}(x)e^{\alpha x}, \qquad i = 1, \ldots, n,$$

where each $M_i^{*(\beta+p)}$ is a polynomial in x of degree no higher than $\beta + p$, and p is the highest degree of an elementary divisor of the form $(\lambda - \alpha)^{\delta}$ of the characteristic matrix $\lambda E - A$.

Finally we consider the implications of the above theory for a single equation of degree n. Suppose α is a zero of order $p \geqslant 0$ of the polynomial (6.17) introduced on p. 143. Then the equation

$$\frac{d^n y}{dx^n} = a_{n-1}\frac{d^{n-1}y}{dx^{n-1}} + a_{n-2}\frac{d^{n-2}y}{dx^{n-2}} + \cdots + a_1\frac{dy}{dx} + a_0 y + Cx^{\beta}e^{\alpha x} \quad (6.21)$$

---

[11] It would be incorrect to think that the system (6.7) must have a particular solution of the form $x^{\beta}M_i^{*(p)}(x)e^{\alpha x}$, since besides the given group of equations corresponding to $\lambda_s = \alpha$, there are other groups corresponding to $\lambda_s \neq \alpha$.

has a particular solution of the form

$$y(x) = M^{(\beta+p)}(x)e^{\alpha x}, \qquad (6.22)$$

where $M^{(\beta+p)}(x)$ is a polynomial of degree no higher than $\beta + p$. By subtracting an appropriate particular solution (6.15) of the homogeneous equation from (6.22), we can obtain a particular solution of the nonhomogeneous equation of the form

$$y(x) = M^{(\beta)}(x)x^p e^{\alpha x}.$$

Problem 1. Use the result of Prob. 5, Sec. 46 to find a particular solution of equation (6.18).

Problem 2. Prove that the general solution of the system (6.1) is of the form

$$y = e^{xA}C + \int_{x_0}^{x} e^{(x-s)A}f(s)\,ds,$$

where C is an arbitrary constant vector. Prove that the solution of (6.1) satisfying the initial condition $y(x_0) = y^0$ is

$$y = e^{(x-x_0)A}y^0 + \int_{x_0}^{x} e^{(x-s)A}f(s)\,ds.$$

Problem 3. Using the last result and Prob. 3, Sec. 19, prove that if every solution of the system (6.11) is bounded as $x \to \infty$, then so is every solution of the system

$$y' = [A + B(x)]y + f(x),$$

where

$$\int_{x_0}^{\infty} \|B(x)\|\,dx < \infty, \qquad \int_{x_0}^{\infty} \|f(x)\|\,dx < \infty$$

[the matrix $B(x)$ and the vector $f(x)$ are continuous]. Moreover, prove that if every solution of the first system approaches zero as $x \to \infty$, then so does every solution of the second system. Find conditions under which analogous assertions hold for linear systems with variable coefficients.[12]

48. Reduction of the Equation $y' = \dfrac{ax + by}{cx + dy}$ to Canonical Form

We now consider the equation

$$\frac{dy}{dx} = \frac{ax + by}{cx + dy}, \qquad (6.23)$$

with real coefficients a, b, c and d. Equation (6.23) is equivalent to the system

$$\frac{dx}{dt} = cx + dy, \qquad \frac{dy}{dt} = ax + by \qquad (6.24)$$

---

[12] For a number of results along these lines, see R. Bellman, *op. cit.*

involving the auxiliary variable t. Depending on the elementary divisors of the λ-matrix

$$\left\| \begin{array}{cc} \lambda - c & -d \\ -a & \lambda - b \end{array} \right\|, \tag{6.25}$$

three cases can arise in reducing (6.24) to canonical form.

Case 1. Suppose there are two real elementary divisors $\lambda - \lambda_1$ and $\lambda - \lambda_2$. In this case, according to Sec. 42, there exists a nonsingular linear transformation

$$x^* = k_{11}x + k_{12}y, \qquad y^* = k_{21}x + k_{22}y \tag{6.26}$$

with real coefficients which reduces (6.24) to the form

$$\frac{dx^*}{dt} = \lambda_1 x^*, \qquad \frac{dy^*}{dt} = \lambda_2 y^*. \tag{6.27}$$

If the determinant

$$\left| \begin{array}{cc} c & d \\ a & b \end{array} \right| \tag{6.28}$$

is nonzero, then λ_1 and λ_2 are both nonzero, and equation (6.23) takes the form

$$\frac{dy^*}{dx^*} = \frac{\lambda_2 y^*}{\lambda_1 x^*}$$

after making the transformation (6.26).

Case 2. Next suppose the matrix (6.25) *has two complex elementary divisors $\lambda - \lambda_1$ and $\lambda - \lambda_2$.* Then λ_1 and λ_2 are complex conjugates, since the coefficients a, b, c and d are real, and the transformation (6.26) again reduces the system (6.24) to the form (6.27). The coefficients k_{11} and k_{12} are now complex conjugates of the coefficients k_{21} and k_{22}, respectively. In fact, since $\lambda_2 = \bar{\lambda}_1$ (the overbar denotes the complex conjugate), it follows from the first of the equations (6.27) that

$$\frac{d\bar{x}^*}{dt} = \bar{\lambda}_1 \bar{x}^* = \lambda_2 \bar{x}^*,$$

and hence we can choose

$$y^* = \bar{x}^*.$$

Introducing the notation

$$\lambda_1 = \alpha + i\beta \qquad (\beta \neq 0),$$
$$k_{11} = \gamma_1 + i\delta_1, \qquad k_{12} = \gamma_2 + i\delta_2,$$
$$\xi = \gamma_1 x + \gamma_2 y, \qquad \eta = \delta_1 x + \delta_2 y,$$

and taking real and imaginary parts in (6.27), we find that

$$\frac{d\xi}{dt} = \alpha\xi - \beta\eta, \qquad \frac{d\eta}{dt} = \beta\xi + \alpha\eta,$$

which implies

$$\frac{d\eta}{d\xi} = \frac{\beta\xi + \alpha\eta}{\alpha\xi - \beta\eta}.$$

It should be noted that the linear transformation carrying x and y into ξ and η is nonsingular, since otherwise we would have

$$\begin{vmatrix} k_{11} & k_{12} \\ k_{21} & k_{22} \end{vmatrix} = 0.$$

Case 3. Finally let the matrix (6.20) *have one elementary divisor* $(\lambda - \lambda_1)^2$. Then, according to Sec. 42, there exists a nonsingular linear transformation (6.26) with real coefficients which reduces (6.24) to the form

$$\frac{dx^*}{dt} = \lambda_1 x^*, \qquad \frac{dy^*}{dt} = \varepsilon x^* + \lambda_1 y^*, \qquad (6.29)$$

where ε is an arbitrary nonzero number. If the determinant (6.28) is nonzero, then $\lambda_1 \neq 0$. Moreover λ_1 is real, since the coefficients a, b, c and d are real. Choosing ε to be real, we find that the coefficients k_{ij} are also real (recall the considerations of Sec. 42). For example, suppose $\varepsilon = \lambda_1$. Then (6.29) implies

$$\frac{dy^*}{dx^*} = \frac{\lambda_1 x^* + \lambda_1 y^*}{\lambda_1 x^*} = \frac{x^* + y^*}{x^*}.$$

Problem 1. Suppose equation (6.23) has a focus at the origin. Under what conditions (involving the coefficients a, b, c and d) do the spirals wind in the counterclockwise direction as they approach the singular point? Under what conditions do they wind in the clockwise direction? Solve the analogous problem for a node (of the type shown in Figure 15, p. 69).

Problem 2. Suppose equation (6.23) has a saddle point or a node at the origin. Along what directions do the integral curves approach the origin in the case of a saddle point? Along what direction does the infinite family of integral curves approach the origin in the case of a node? If the origin is a center, locate the principal axes of the ellipses. Which axis is larger?

49. Stability of Solutions

Suppose the initial data are specified for $x = x_0$. Then a solution

$$y_i = y_i^0(x), \qquad i = 1, \ldots, n \qquad (6.30)$$

of the system[13]

$$\frac{dy_i}{dx} = f_i(x, y_1, \ldots, y_n), \qquad i = 1, \ldots, n \qquad (6.31)$$

is said to be *stable* (in the sense of Lyapunov) as $x \to +\infty$ if given any $\varepsilon > 0$, there is an $\eta(\varepsilon) > 0$ such that an arbitrary solution $y_i(x)$, $i = 1, \ldots, n$ of (6.31) satisfies the inequality

$$|y_i(x) - y_i^0(x)| < \varepsilon$$

for all $x \geqslant x_0$, provided that

$$|y_i(x_0) - y_i^0(x_0)| < \eta(\varepsilon).$$

If in addition

$$\lim_{x \to +\infty} [y_i(x) - y_i^0(x)] = 0$$

for sufficiently small $|y_i(x_0) - y_i^0(x_0)|$, $i = 1, \ldots, n$, we say that the solution (6.30) is *asymptotically stable* (in the sense of Lyapunov) as $x \to +\infty$. Here, of course, it is assumed that the functions $y_i^0(x)$ are defined for all $x \geqslant x_0$ and that the system (6.31) is defined in some "neighborhood" of the curve $y_i = y_i^0(x)$, $i = 1, \ldots, n$ of the form

$$x \geqslant x_0, \quad |y_i - y_i^0(x)| \leqslant M, \qquad i = 1, \ldots, n.$$

Obviously, we need only consider the case

$$y_i(x) \equiv 0, \qquad i = 1, \ldots, n,$$

since the $y_i(x)$ can always be replaced by new unknown functions $y_i(x) - y_i^0(x)$. Finally, we assume that x and all the functions f_i, y_i are real.

LEMMA (*Lyapunov's lemma*).[14] *Suppose that for some $\varepsilon_0 > 0$ the functions $f_i(x, y_1 \ldots y_n)$ are defined and continuous for $x_0 \leqslant x < \infty$, $-\varepsilon_0 \leqslant y_i \leqslant \varepsilon_0$, $i = 1, \ldots, n$, and satisfy the conditions*

$$f_i(x, 0, \ldots, 0) \equiv 0, \qquad i = 1, \ldots, n.$$

Suppose further that for the same values of y_i there exists a continuously differentiable "Lyapunov function" $V(y_1, \ldots, y_n) \geqslant 0$ equal to zero only at the origin, such that

$$\sum_{j=1}^{n} \frac{\partial V}{\partial y_j} f_j \leqslant 0. \qquad (6.32)$$

---

[13] We now return to the general system (4.4).

[14] The pioneering work on stability of solutions of differential equations is due to A. M. Lyapunov, *Problème général de la stabilité du mouvement*, Ann. Fac. Sci. Univ. Toulouse, **9**, 203 (1907), reprinted in Annals of Mathematics Studies, No. 17, Princeton University Press, Princeton, N.J. (1947).

Then the null solution $y_i(x) \equiv 0$, $i = 1, \ldots, n$ *of the system* (6.31) *is stable. Moreover, the null solution is asymptotically stable if*

$$\sum_{j=1}^{n} \frac{\partial V}{\partial y_j} f_j < -W(y_1, \ldots, y_n) < 0 \qquad (6.33)$$

for the same values of y_i, *where* $W(y_1, \ldots, y_n) \geqslant 0$ *is a continuous function equal to zero only at the origin.*

Proof. Let $0 < \varepsilon \leqslant \varepsilon_0$ and let K_ε denote the surface of the n-dimensional cube

$$-\varepsilon \leqslant y_i \leqslant \varepsilon, \qquad i = 1, \ldots, n.$$

Moreover, let

$$V_\varepsilon = \min_{K_\varepsilon} V,$$

where clearly $V_\varepsilon > 0$. Then choose $\eta > 0$, $\eta < \varepsilon$ so small that $V < V_\varepsilon$ on K_η and everywhere inside K_η. Such a value of η exists, since V is continuous and vanishes at the origin. Since V is a composite function of x along any integral curve l, it follows from (6.31) that

$$\frac{dV}{dx} = \sum_{j=1}^{n} \frac{\partial V}{\partial y_j} \frac{dy_j}{dx} = \sum_{j=1}^{n} \frac{\partial V}{\partial y_j} f_j$$

along l. Thus the condition (6.32) means that V cannot increase along l as x increases. Suppose that as x increases, l starts at a point $x = x_0$ inside K_η and subsequently reaches K_ε for the first time at some point $x = x_1$. Then

$$V|_{x=x_0} < V_\varepsilon \leqslant V|_{x=x_1}$$

along l, contrary to the fact that V cannot increase along l. This contradiction shows that no integral curve which begins at a point $x = x_0$ inside K can ever reach K_ε as $x \to +\infty$, thereby proving the stability.

If the stronger condition (6.33) is satisfied, then in addition to being a nonincreasing function of x along l, V approaches zero along l as $x \to +\infty$. In fact, suppose V does not approach zero as $x \to +\infty$. Then V exceeds some positive constant everywhere on l, i.e., V lies entirely outside some sufficiently small cube K_δ It follows from (6.33) that

$$\frac{dV}{dx} \leqslant -W \leqslant -\alpha < 0$$

along l, since $W \geqslant \alpha > 0$ ($\alpha = $ const) outside K_δ. Integrating this inequality, we find that

$$V(x) \leqslant V(x_0) - \alpha(x - x_0),$$

where the right-hand side approaches $-\infty$ as $x \to +\infty$, contrary to the definition of V. Therefore $V \to 0$ along l as $x \to +\infty$. But then every integral curve l beginning at a point x_0 inside K_η not only stays inside K_ε as $x \to +\infty$, but actually approaches the origin, thereby proving the asymptotic stability.

Remark 1. We emphasize again that the left-hand side of (6.32) or (6.33) is the derivative dV/dx of the function V evaluated along integral curves of the system (6.31).

Remark 2. Lyapunov's lemma has a simple geometric interpretation, as illustrated by the following special case: Let $n = 2$, and suppose the curves $V = C (C = \text{const})$ are closed and contain the origin. Moreover, suppose $C < C'$ (see Figure 27). Then the condition (6.32) means that no integral curve sharing a point with the curve $V = C$ can leave the domain bounded by $V = C$. It follows that the null solution $y_1 \equiv 0$, $y_2 \equiv 0$ (the origin of the xy-plane) is stable. If the stronger condition (6.33) holds, then the integral curves cross the curve $V = C$ from the outside to the inside, since

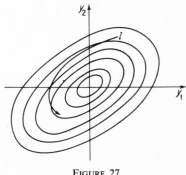

FIGURE 27

$$\frac{dV}{dx}\bigg|_{V=C} \leqslant -W|_{V=C} < -\beta \qquad (\beta > 0).$$

Moreover, according to the lemma, $V \to 0$ as $x \to +\infty$. Therefore every integral curve approaches the origin as $x \to +\infty$, i.e., the null solution is asymptotically stable.

THEOREM. *Suppose the functions f_i figuring in the system (6.31) satisfy the conditions of Lyapunov's lemma and are of the form*

$$f_i(x, y_1, \ldots, y_n) = \sum_{j=1}^{n} a_{ij} y_j + F_i(x, y_1, \ldots, y_n),$$

where the a_{ij} are constant, and the real parts of all roots λ of the characteristic equation

$$\det(\lambda E - A) = 0 \qquad (6.34)$$

are negative. Suppose further that

$$|F_i(x, y_1, \ldots, y_n)| \leqslant M[|y_1|^{1+\alpha} + \cdots |y_n|^{1+\alpha}] \qquad (6.35)$$

for all $x \geqslant x_0$ and sufficiently small $|y_i|$, where the functions F_i are continuous (in all their arguments jointly) and M, α are positive constants.

Then the null solution

$$y_i(x) \equiv 0, \qquad i = 1, \ldots, n$$

of (6.31) *is asymptotically stable as* $x \to +\infty$.

Proof. It should be noted that the conditions of the theorem are satisfied if the functions f_i are independent of x and have continuous first and second derivatives in some neighborhood of the origin, and if the real parts of all roots of equation (6.34) are negative. In fact, we can then expand the f_i in Taylor series, obtaining[15]

$$f_i(x, y_1, \ldots, y_n) = \sum_{j=1}^{n} a_{ij} y_j + O\left(\sum_{j=1}^{n} y_i^2 \right).$$

Turning to the proof itself, we first consider the particularly simple equation

$$\frac{dy}{dx} = -ay \qquad (a = \text{const} > 0).$$

In this case we have

$$\frac{d(y^2)}{dx} = 2y \frac{dy}{dx} = -2ay^2,$$

and hence (6.33) and all the other conditions of Lyapunov's lemma are satisfied by setting

$$V(y) = y^2, \qquad W(y) = 2ay^2.$$

This suggests that we subject the system (6.31) to a nonsingular linear transformation reducing the "linear part" of the system to canonical form, and then take the Lyapunov function to be the sum of the squares of the absolute values of the new unknown functions. Pursuing this idea in detail, we make a nonsingular linear transformation

$$y_i = \sum_{j=1}^{n} c_{ij} z_j$$

reducing the system

$$\frac{dy_i}{dx} = \sum_{j=1}^{n} a_{ij} y_j, \qquad i = 1, \ldots, n$$

to canonical form, confining ourselves to values of the variables z_i (in general, complex) for which the corresponding values of the y_i are real.

Next we write

$$V(y_1, \ldots, y_n) = \sum_{i=1}^{n} |z_i|^2 = \sum_{i=1}^{n} z_i \bar{z}_i,$$

where, of course, every z_i on the right is regarded as replaced by its

---

[15] For the meaning of the symbol O, see footnote 37, p. 67.

expression in terms of the y_i. It follows that

$$\frac{dV}{dx} = \sum_{i=1}^n \left(\frac{dz_i}{dx} \bar{z}_i + z_i \frac{d\bar{z}_i}{dx} \right).$$

To calculate these derivatives, we consider the group of canonical equations corresponding to some elementary divisor, say $(\lambda - \lambda_1)^{n_1}$, of the matrix $\lambda E - A$:

$$\frac{dz_1}{dx} = \lambda_1 z_1 \qquad\qquad\qquad + F_1^*(x, z_1, \dots, z_n),$$

$$\frac{dz_2}{dx} = \beta_1 z_1 + \lambda_1 z_2 \qquad\qquad + F_2^*(x, z_1, \dots, z_n),$$

$$\cdots \cdots \cdots \cdots \cdots \cdots \cdots \cdots \cdots$$

$$\frac{dz_{n_1}}{dx} = \qquad\qquad \beta_1 z_{n_1-1} + \lambda_1 z_{n_1} + F_{n_1}^*(x, z_1, \dots, z_n).$$

$$(6.36)$$

Here λ_1 is one of the roots of equation (6.34), and β_1 is an arbitrary nonzero number. If the functions F_i satisfy the condition (6.35), then the functions F_1^* satisfy a condition of the same form

$$|F_i^*(x, z_1, \dots, z_n)| \leqslant M^*[|z_1|^{1+\alpha} + \cdots + |z_n|^{1+\alpha}], \qquad (6.37)$$

where M^* is a new constant. In fact, the F_1^* are linear combinations of the F_i with constant coefficients, and hence there are constants M_1 and M_2 such that

$$|F_i^*(x, z_1, \dots, z_n)| \leqslant M_1 \sum_{j=1}^n |F_j(x, y_1, \dots, y_n)|$$
$$\leqslant M_2[|y_1|^{1+\alpha} + \cdots + |y_n|^{1+\alpha}],$$

where every y_i is a linear combination of the z_i with constant coefficients. Therefore

$$|y_i|^{1+\alpha} \leqslant M_3(\max |z_j|)^{1+\alpha} \leqslant M_3[|z_1|^{1+\alpha} + \cdots + |z_n|^{1+\alpha}]$$

for every i, where M_3 is a suitable constant and $\max |z_j|$ denotes the largest of the numbers $|z_1|, \dots, |z_n|$, and (6.37) follows at once.

Using the equations (6.36), we find that

$$\frac{d|z_i|^2}{dx} = \bar{z}_i \frac{dz_i}{dx} + z_i \frac{d\bar{z}_i}{dx} \qquad\qquad (6.38)$$
$$= \lambda_1 z_i \bar{z}_i + \bar{\lambda}_1 \bar{z}_i z_i + \beta_1 z_{i-1} \bar{z}_i + \bar{\beta}_1 \bar{z}_{i-1} z_i + \bar{z}_i F_i^* + z_i \bar{F}_i^*$$

for $i > 1$. For $i = 1$, we have the same equation without the terms in β_1 and $\bar{\beta}_1$. Writing

$$\text{Re } \lambda_1 = a_1$$

and using the estimate (6.37), we deduce from (6.38) that

$$\frac{d\,|z_i|^2}{dx} \leqslant 2a_1\,|z_i|^2 + |\beta_1|[|z_i|^2 + |z_{i-1}|^2]$$
$$+ 2M^*[|z_1|^{1+\alpha} + \cdots + |z_n|^{1+\alpha}]\max|z_j|, \tag{6.39}$$
$$i = 2, \ldots, n_1,$$

$$\frac{d\,|z_1|^2}{dx} \leqslant 2a_1\,|z_1|^2 + 2M^*[|z_1|^{1+\alpha} + \cdots + |z_n|^{1+\alpha}]\max|z_j|. \tag{6.39'}$$

But clearly

$$[|z_1|^{1+\alpha} + \cdots + |z_n|^{1+\alpha}]\max|z_j| \leqslant n(\max|z_j|^2)^{1+(\alpha/2)} \leqslant nV^{1+(\alpha/2)}.$$

Therefore, summing the inequalities (6.39) and (6.39') over all i from 1 to n, and letting $-a$ denote the largest real part of all the roots λ_i (which is negative by hypothesis), we find that

$$\frac{dV}{dx} = \frac{d}{dx}\sum_{i=1}^{n}|z_i|^2 \leqslant -2aV + 2BV + 2M^*n^2V^{1+(\alpha/2)},$$

where B is the largest of the numbers $|\beta_s|$. Since the β_i can be made arbitrarily small in absolute value, provided only that they are nonzero, we can choose them so that

$$B < \frac{a}{4}.$$

Moreover, let all the y_i and hence all the z_i be so small in absolute value that

$$V^{\alpha/2} < \frac{a}{4M^*n^2}.$$

Then

$$\frac{dV}{dx} \leqslant -aV,$$

and setting $W = aV$, we see that (6.33) and the other conditions for applying Lyapunov's lemma are satisfied. Therefore the null solution of (6.31) is asymptotically stable as $x \to +\infty$, as asserted.

Remark. The stability property does not depend on the choice of x_0 if the system (6.31) has a unique solution passing through each point of the curve $y_i = y_i^0(x)$, $i = 1, \ldots, n$, $x \geqslant x_0$. In other words, the solution $y_i = y_i^0(x)$, $i = 1, \ldots, n$ of the system (6.31) is stable for initial data specified at any fixed point $x_0' \geqslant x_0$ if and only if it is stable for initial data specified at x_0. This is an immediate consequence of the fact that solutions depend continuously on the initial data in the finite interval $x_0 \leqslant x \leqslant x_0'$ (see Secs. 19 and 29, in particular the remark on p. 58).

Problem 1. The lemma proved in the text might be called the *stability lemma*. Prove the following *instability lemma*, also due to Lyapunov: Suppose all the conditions of the stability lemma are satisfied, except that it is now required that V take positive values in every neighborhood of the origin and that

$$\sum_{j=1}^{n} \frac{\partial V}{\partial y_j} f_j \geqslant U(y_1, \ldots, y_n),$$

where U is a continuous function which is positive wherever V is positive. Then the null solution of the system (6.31) is *unstable*.

Give examples showing that both lemmas fail if the function V is allowed to depend on x. However, give further requirements guaranteeing that the lemmas remain valid.

Problem 2. Using the instability lemma, prove that if all the conditions of the theorem on p. 153 are satisfied, except that at least one root of equation (6.34) has a positive real part, then the null solution is unstable.

Hint. Set

$$V(y_1, \ldots, y_n) = \sum_{i=1}^{m} |z_i|^2 - \sum_{i=m+1}^{n} |z_i|^2,$$

where $z_1, \ldots, z_n$ are the "canonical variables" corresponding to roots with positive real parts.

Problem 3. Construct an example of a system (6.31) with only one stable solution, which nevertheless has a unique solution which satisfies arbitrary initial data and is bounded for all x.

Problem 4. Show that even if the solution of the system (6.31) satisfying arbitrary initial data converges to the null solution $y_i(x) \equiv 0$, $i = 1, \ldots, n$ as $x \to +\infty$, it is still not necessary that the null solution be stable (give an example). Suppose that in addition it is known that the null solution is stable. Then must all solutions with "sufficiently close" initial data also be stable? Analyze the cases $n = 1$ and $n > 1$ separately.

Problem 5. Show that if all solutions satisfying the condition

$$|y_i(x_0)| < M, \qquad i = 1, \ldots, n \tag{6.40}$$

converge uniformly as $x \to +\infty$ to the null solution, $y_i \equiv 0$, $i = 1, \ldots, n$, then all solutions satisfying (6.40) are stable.

Problem 6. Prove that in the theorem on p. 153 and in Prob. 2, we can replace the condition (6.35) by the weaker condition

$$F_i(x, y_1, \ldots, y_n) = o(|y_1| + \cdots + |y_n|),$$

i.e.,

$$\frac{F_i(x, y_1, \ldots, y_n)}{|y_1| + \cdots + |y_n|} \to 0, \qquad i = 1, \ldots, n$$

(uniformly in x) as all the $y_i \to 0$.

Problem 7. Let $n = 1$, and suppose two solutions satisfying initial data $y(x_0) = y^{01}$ and $y(x_0) = y^{02} > y^{01}$ converge asymptotically to the same (finite) limit as $x \to +\infty$. Under these conditions, prove that if the initial data uniquely determine the solution, then every solution such that $y^{01} < y(x_0) < y^{02}$ is stable.

Problem 8. Find a necessary and sufficient condition for stability of the null solution of a system of homogeneous linear differential equations with constant coefficients.

Problem 9. Prove that a necessary and sufficient condition for stability of the null solution of a system of homogeneous linear differential equations with continuous coefficients is that every solution of the system be bounded.

Problem 10. Let

$$\frac{dy}{dx} = A(x)y \tag{6.41}$$

be a linear system in matrix form with continuous periodic coefficients, i.e., such that $A(x + a) = A(x)$ where $a = \text{const} > 0$. Let $Y(x)$ be the fundamental matrix of the system (see Prob. 1, Sec. 33). Prove that $Y(x + a) \equiv TY(x)$, where T is some constant matrix. How is this matrix affected by changing the fundamental matrix? Find conditions for stability and asymptotic stability of solutions of (6.41), expressed in terms of the properties of the matrix T.

Problem 11 (M. A. Krasnoselski and S. G. Krein). Suppose the functions f_i figuring in the system (6.31) are defined and continuous for all $x_1 \leqslant x < \infty$, $-\infty < y_i < \infty$, $i = 1, \ldots, n$. Moreover, suppose there exists a continuously differentiable function $V(y_1, \ldots, y_n)$, defined for all y_i, such that

$$\lim_{|y_1| + \cdots + |y_n| \to \infty} V = \infty,$$

and a continuous function $\Phi(V) > 0$ [$\min V(y) \leqslant V < \infty$] such that

$$\int_0^\infty \frac{dV}{\Phi(V)} = \infty, \qquad \sum_{j=1}^n \frac{\partial V}{\partial y_j} f_j \leqslant \Phi(V).$$

Prove that any solution defined for $x = x_0$, $x_1 \leqslant x_0 < \infty$ can be continued in the direction of increasing x onto the interval $x_0 \leqslant x < \infty$.

Next suppose the functions f_i figuring in the system (6.31) are defined and continuous for all $x_1 \leqslant x < \infty$, $-\varepsilon_0 \leqslant y_i \leqslant \varepsilon_0$, $i = 1, \ldots, n$, $\varepsilon_0 > 0$, where all the $f_i(x, 0, \ldots, 0) \equiv 0$. State an analogous theorem giving conditions making it impossible for an integral curve beginning for $x = x_0 (x_1 \leqslant x_0 < \infty)$ at a point other than the origin of y-space to arrive at the origin for some finite value of x. (At the same time, the theorem gives sufficient conditions for the uniqueness of solutions beginning at points of the axis $y_1 = \cdots = y_n = 0$.) Try to combine this theorem and the theorem in the preceding paragraph into a single theorem on the impossibility of an integral curve leaving some domain of y-space for a finite value of x.

50. A Physical Example

Consider a particle of mass $m > 0$ moving along the x-axis, subject to a resistive force

$$-a \frac{dx}{dt} \qquad (a = \text{const} \geqslant 0)$$

proportional to its velocity (due to the surrounding medium, e.g., a liquid or a gas), a restoring force

$$-bx \qquad (b = \text{const} > 0)$$

applied, say, by a spring obeying Hooke's law,[16] and an external periodic force which at time t equals

$$A \cos \omega t,$$

where A and ω are real constants ($\omega > 0$). Then the differential equation describing the motion of the particle takes the form

$$m \frac{d^2x}{dt^2} + a \frac{dx}{dt} + bx = A \cos \omega t. \qquad (6.42)$$

First we study the case where $A = 0$. This is the case of no restoring force, corresponding to "free oscillations" of the particle. Assuming that the roots λ_1 and λ_2 of the characteristic equation

$$m\lambda^2 + a\lambda + b = 0 \qquad (6.43)$$

are distinct, i.e., that

$$\lambda_{1,2} = -\frac{a}{2m} \pm \sqrt{\frac{a^2}{4m^2} - \frac{b}{m}},$$

we find that the general solution of the homogeneous equation

$$m \frac{d^2x}{dt^2} + a \frac{dx}{dt} + b = 0 \qquad (6.44)$$

is given by the formula

$$x = C_1 e^{\lambda_1 t} + C_2 e^{\lambda_2 t}. \qquad (6.45)$$

If $a > 0$, the real parts of λ_1 and λ_2 are negative, and hence, by inspection, every solution of equation (6.44) converges to zero as $t \to +\infty$. By the same token, every solution of the system

$$\frac{dx}{dt} = x_1,$$

$$m \frac{dx_1}{dt} = -bx - ax_1 \qquad (6.46)$$

----

[16] According to Hooke's law, the elastic force exerted by the spring acts in the direction of the equilibrium position of the particle, and is proportional to the displacement of the particle from its equilibrium position.

corresponding to (6.44), converges to the null solution $x(t) \equiv 0$, $x_1(t) \equiv 0$ as $t \to +\infty$. This can also be inferred from the theorem on p. 153.

On the other hand, suppose $a = 0$. Then every real solution of (6.44) is given by the formula

$$x(t) = C_1 \sin \sqrt{b/m}\ t + C_2 \cos \sqrt{b/m}\ t = C \sin (\sqrt{b/m}\ t + \theta),$$

where

$$C = \sqrt{C_1^2 + C_2^2}, \qquad C_1 = C \cos \theta, \qquad C_2 = C \sin \theta.$$

It follows that

$$x_1(t) = \frac{dx}{dt} = C \sqrt{b/m} \cos (\sqrt{b/m}\ t + \theta).$$

Therefore the point $(x(t), x_1(t))$ moves in an elliptical orbit in the xx_1-plane, where the axes of the ellipse lie along the coordinate axes. The ratio of the semiminor axis to the semimajor axis is the same for every ellipse, i.e., $\sqrt{b/m}$, and the origin is clearly a center for the system (6.46). Moreover, the particle oscillates along the x-axis, with period $2\pi\sqrt{m/b}$ which is the same for every solution of (6.44).

Next we consider the motion for $a > 0$ in somewhat more detail, distinguishing three cases:

Case 1. If $a^2 > 4bm$, both roots of the characteristic equation (6.43) are real and negative. Clearly, neither the function $x(t)$ given by (6.45) nor its derivative can vanish for more than one value of t. In particular, $x(t)$ has no more than one maximum or minimum. In this case, the origin of the xx_1-plane is a node of the system (6.44), as in Figures 12 and 13, p. 68.

Case 2. If $a^2 = 4bm$, the solution of (6.44) is given by

$$x = e^{-at/2m}(C_1 + C_2 t).$$

As before, neither of the functions $x(t)$ and $x_1(t)$ can vanish for more than one value of t. The origin of the xx_1-plane is again a node of the system (6.44), of the type shown in Figure 15, p. 69.

Case 3. If $a^2 < 4bm$, the roots of the characteristic equation (6.43) are conjugate complex numbers, of the form

$$\lambda_{1,2} = -\alpha + i\beta \qquad \left(\alpha = \frac{a}{2m} > 0, \beta > 0\right),$$

and the real solutions of equation (6.44) are given by

$$x = e^{-\alpha t}(C_1 \sin \beta t + C_2 \cos \beta t) = Ce^{-\alpha t} \sin (\beta t + \theta).$$

The particle oscillates along the x-axis with exponentially decaying amplitude $Ce^{-\alpha t}$, and with period $2\pi/\beta$ which is the same for every solution of (6.44).

Finally, we analyze the case where $A \neq 0$ in equation (6.42). It is now more convenient to start from the equation

$$m\frac{dz^2}{dt^2} + a\frac{dz}{dt} + bz = Ae^{i\omega t} \qquad (6.47)$$

instead of (6.42), noting that the real part of every solution of (6.47) satisfies (6.42), while conversely every solution of (6.42) is the real part of some solution of equation (6.47) [why?].

If both roots of (6.43) are different from $i\omega$, the general solution of (6.47) is given by

$$z = C_1 e^{\lambda_1 t} + C_2 e^{\lambda_2 t} + \frac{Ae^{i\omega t}}{m(i\omega)^2 + a(i\omega) + b}$$

if $\lambda_1 \neq \lambda_2$, and by

$$z = C_1 e^{\lambda t} + C_2 t e^{\lambda t} + \frac{Ae^{i\omega t}}{m(i\omega)^2 + a(i\omega) + b}$$

if $\lambda_1 = \lambda_2 = \lambda$. The first two terms in these formulas give the general solution of the homogeneous equation (6.44). This solution is bounded as $t \to +\infty$, for arbitrary C_1 and C_2. The second term is a particular solution of equation (6.47), found by the rule given at the end of Sec. 47. If $a > 0$, the first two terms in both formulas approach zero as $t \to +\infty$, and hence the solution of (6.47) approaches

$$\frac{Ae^{i\omega t}}{m(i\omega)^2 + a(i\omega) + b}. \qquad (6.48)$$

Thus the smaller the absolute value of $m(i\omega)^2 + a(i\omega) + b$, the larger the absolute value of the function (6.48), for fixed A.

If $m(i\omega)^2 + a(i\omega) + b = 0$, which can only happen if $a = 0$, the general solution of (6.47) becomes

$$z = C_1 e^{i\omega t} + C_2 e^{-i\omega t} + \frac{Ate^{i\omega t}}{2m(i\omega)}. \qquad (6.49)$$

The first two terms of (6.49) give the general solution of the homogeneous equation (6.44), and remain bounded as $t \to +\infty$. The second term is a particular solution of equation (6.41), found by the method of Sec. 47, and goes to infinity in absolute value as $t \to +\infty$. In other words, the solution of (6.47) is now an oscillatory function whose amplitude increases without limit. In physics this phenomenon is called *resonance* (in the present case, between the free oscillations of the particle and the external force). As just shown, resonance occurs when the period of the free oscillations of the particle coincides with the period of the external force. It is important to anticipate the possibility of resonance occurring in an actual physical system, since the resulting "sympathetic oscillations" may well become large enough to eventually destroy the system.

Problem 1. Analyze in detail the case where $a < 0$, corresponding to oscillations with negative resistance. This occurs in a number of physical systems when energy is supplied to the system from the outside.

Problem 2. Prove that the null solution of the system (6.46) is stable, by studying the time derivative of the energy integral

$$\frac{m}{2}\left(\frac{dx}{dt}\right)^2 + \frac{b}{2}\, x^2.$$

7

AUTONOMOUS SYSTEMS

---

51. General Concepts

If the independent variable does not appear in the right-hand sides of the system (4.4), the system is said to be *autonomous* (or *dynamical*). In the theory of autonomous systems, the independent variable is customarily interpreted as the time, and the system is then written in the form

$$\frac{dx_i}{dt} = f_i(x_1, \ldots, x_n), \qquad i = 1, \ldots, n, \tag{7.1}$$

or more concisely as

$$\frac{dx}{dt} = f(x), \tag{7.2}$$

where

$$x = \begin{Vmatrix} x_1 \\ x_2 \\ \cdot \\ \cdot \\ \cdot \\ x_n \end{Vmatrix}, \qquad f = \begin{Vmatrix} f_1 \\ f_2 \\ \cdot \\ \cdot \\ \cdot \\ f_n \end{Vmatrix}.$$

However, the symbol x will also be used to denote the point with coordinates $x_1, \ldots, x_n$.

For simplicity, we shall assume that the functions f_i appearing in (7.1) are defined for all x and satisfy a Lipschitz condition in all their arguments in every bounded part of space. Then there is a unique solution $x = x(t; x^0)$ of the system (7.1) [equivalently, of the system (7.2)] satisfying the initial condition $x(0) = x^0$ and defined in a neighborhood of the value $t = 0$.[1] This solution can be interpreted as the motion of a variable point x in a space of n dimensions, where x describes a *trajectory* l_{x^0} which depends on the choice of the initial point x^0. Note that a trajectory is not an integral curve, and in fact integral curves of (7.2) lie in the $(n + 1)$-dimensional space of points (x, t).

The velocity vector corresponding to the motion $x = x(t)$ is given by

$$v = \frac{dx}{dt}.$$

Therefore the autonomous system specifies a *velocity field* in x-space, assigning a vector $v = f(x)$ to every point x. A solution of (7.2) corresponds to a motion such that the variable point x has the given velocity $v = f(x)$ at each point of its trajectory. The fact that t does not appear in the right-hand side of (7.2) means that the velocity field is *stationary*, i.e., does not vary in time. Thus the system (7.2) can be interpreted physically as the stationary flow of a gas in x-space, with the solutions describing motions of the gas particles.

As we know, one of two cases can occur when the solution $x(t; x^0)$ is continued in the direction of increasing t (similarly for decreasing t): Either the solution can be continued onto the whole strip $0 \leqslant t < \infty$, or else the point $x(t; x^0)$ "goes off to infinity" for some finite $t = T$ (cf. Remark 3, p. 32) For simplicity, we shall assume that the first case always occurs. It is easy to see that this does not really lead to any loss of generality in studying the trajectories of the system (7.2). In fact consider the autonomous system

$$\frac{dx_i}{dt} = f_i(x_1, \ldots, x_n)\rho(x_1, \ldots, x_n), \qquad i = 1, \ldots, n, \qquad (7.3)$$

where the function ρ satisfies the same requirements as the f_i and is nonvanishing. Then (7.3) has the same trajectories as (7.1), although the trajectories of (7.1) and (7.3) are not traversed with the same velocity. Moreover, for an appropriate choice of ρ, the velocity of the motion given by (7.3) will be uniformly bounded, so that the moving point cannot go off to infinity in finite time. In fact, we need only choose

$$\rho = \left(1 + \sum_{i=1}^{n} f_i^2\right)^{-1/2}.$$

Thus we shall henceforth assume that the solution of (7.4) equal to x^0 for $t = 0$ is defined for all x, x^0 in x-space and for all t ($-\infty < t < \infty$). Let this solution be denoted by

$$x = x(t; x^0), \qquad (7.4)$$

---

[1] Here we write x^0 instead of x_0 to prevent any possible confusion with a component of the vector x.

where, figuratively speaking, $x(t; x^0)$ is the point into which x^0 moves in time t. The function (7.4) has the following properties:

1) $x(t; x^0)$ is continuous in all its arguments;

2) $x(0; x^0) \equiv x^0$;

3) $x[t_2; x(t_1; x^0)] \equiv x(t_1 + t_2; x^0)$.

Property 1 follows from the theorem on continuous dependence of a solution on the initial data, and property 2 is an immediate consequence of the definition of the solution $x(t; x^0)$. To prove property 3 (often called the *group property*), we first note that the absence of t in the right-hand side of (7.2) implies that if $x = \varphi(t)$ is a solution of (7.2), then so is $x = \varphi(t + t_0)$ for arbitrary t_0 (verify this).[2] Therefore

$$x = x[t; x(t_1; x^0)], \qquad x = x(t + t_1; x^0)$$

are both solutions of (7.2). But for $t = 0$, both expressions give the same point $x(t_1; x^0)$, and hence

$$x[t; x(t_1; x^0)] \equiv x(t + t_1; x^0)$$

by the uniqueness theorem for solutions of systems of differential equations. Setting $t = t_2$, we obtain the required property 3. Intuitively, property 3 means that the point into which x^0 moves in time $t_1 + t_2$ can be found in two steps: First find the point P into which x^0 moves in time t_1, and then find the point into which P moves in time t_2. These properties are so important that an autonomous system is often defined as a family of mappings (7.4) of an arbitrary set into itself satisfying properties 1–3, without reference to any underlying differential equations.[3]

It follows from properties 2 and 3 that each of the two mappings

$$x = x(t; x^0), \qquad x = x(-t; x^0)$$

of x-space into itself (where t has any fixed value) is the inverse of the other. In fact,

$$x[-t; x(t; x^0)] = x(-t + t; x^0) = x(0; x^0) = x^0,$$

and similarly,

$$x[t; x(-t; x^0)] = x(t - t; x^0) = x(0; x^0) = x^0.$$

It should also be noted that if two trajectories have a common point, then they coincide and the corresponding solutions differ only by a constant time shift. In fact, if

$$x(t_1; x^1) = x(t_2; x^2),$$

---

[2] The second solution corresponds to the same trajectory l as the first, except that the motion along l is shifted ahead by time t_0.

[3] It is assumed of course that continuity of the mappings on the given set can be suitably defined.

then
$$x(t; x^1) \equiv x[t + (t_2 - t_1); x^2]$$
by the uniqueness theorem, since the two solutions coincide for $t = t_1$.

Remark. Second-order systems of the form

$$m_i \frac{d^2 x_i}{dt^2} = f_i\left(x_1, \ldots, x_n, \frac{dx_1}{dt}, \ldots, \frac{dx_n}{dt}\right), \qquad i = 1, \ldots, n \quad (7.5)$$

are encountered in studying mechanical systems with n degrees of freedom. After introducing new variables

$$m_i \frac{dx_i}{dt} = p_i$$

(called *momenta*), we can write (7.5) in the form

$$\frac{dx_i}{dt} = \frac{p_i}{m_i},$$

$$\frac{dp_i}{dt} = f_i\left(x_1, \ldots, x_n, \frac{p_1}{m_1}, \ldots, \frac{p_n}{m_n}\right),$$

where $i = 1, \ldots, n$. This is an autonomous system in the $2n$-dimensional space of points $(x_1, \ldots, x_n, p_1, \ldots, p_n)$ called *phase space*.

Problem. Suppose we are given a function (7.4) in the n-dimensional space of points $x = (x_1, \ldots, x_n)$ which satisfies properties 1–3 and is continuously differentiable with respect to t. Prove that this function is the solution of an autonomous system of the form (7.2) with a continuous right-hand side.

52. Classification of Trajectories

THEOREM. *A solution $x(t)$ of the autonomous system (7.2) must be of one of the following three types*:

1) *Nonperiodic, which means that $x(t_1) \neq x(t_2)$ if $t_1 \neq t_2$;*

2) *Periodic, which means that $x(t + T) \equiv x(t)$ for some positive constant T (called a period) but $x(t_1) \neq x(t_2)$ if $0 \leqslant t_1 < t_2 < T$;*

3) *Constant, i.e., $x(t) \equiv x^0$.*

Proof. Suppose the solution $x(t)$ is not of type 1, i.e., suppose there are times t_1 and t_2 ($t_1 \neq t_2$) such that $x(t_1) = x(t_2)$. Writing $\tau = t_2 - t_1$, we find that

$$x(t + \tau) \equiv x(t), \qquad (7.6)$$

since the solutions $x(t + \tau)$ and $x(t)$ agree for $t = t_1$. Consider the set K of all numbers τ satisfying (7.6). It is easy to see that if τ belongs to K, so does $-\tau$ [replace t by $t - \tau$ in (7.6)]. Moreover, if τ_1 and τ_2 belong to K, so does $\tau_1 + \tau_2$, since

$$x(t + \tau_1 + \tau_2) \equiv x(t + \tau_1) \equiv x(t),$$

and hence so does $\tau_1 - \tau_2$ (why?).[4]

There are now two possibilities:

a) The set K contains a smallest positive number T. Then $x(t + T) \equiv x(t)$, but $x(t_1) \neq x(t_2)$ if $0 \leqslant t_1 < t_2 < T$, i.e., the solution is periodic and we have a solution of type 2.[5]

b) There is no smallest positive number in K. Then K contains arbitrarily small positive numbers (why?). Hence there is a sequence of positive numbers τ_n belonging to K such that $\tau_n \to 0+$ as $n \to \infty$. But then

$$t - \left[\frac{t}{\tau_n}\right]\tau_n \to 0 \qquad \text{as} \quad n \to \infty,$$

where $[\gamma]$ denotes the integral part of γ, and therefore

$$x(t) = x\left(t - \left[\frac{t}{\tau_n}\right]\tau_n\right) \to x(0) \qquad \text{as} \quad n \to \infty,$$

since the solution $x(t)$ is continuous. In other words, $x(t) \equiv x(0)$, and we have a solution of type 3.[6] The proof is now complete.

DEFINITION. *A trajectory is said to be open if it corresponds to a nonperiodic solution and closed if it corresponds to a periodic solution. A closed trajectory is also called a cycle. A trajectory corresponding to a constant solution is called a critical point.*

53. Limiting Behavior of Trajectories: The General Case

Let $x(t)$ be any solution of the system (7.2), and let l be the corresponding trajectory. A point $\bar{x}$ is said to be a *limit point of the solution* $x(t)$ [or of the trajectory l] *as* $t \to +\infty$ if there is a sequence of times $t_n \to +\infty$ such that $x(t_n) \to \bar{x}$. The set of all such points is called the *limit set of* $x(t)$ *as* $t \to +\infty$. The concepts of a *limit point* and a *limit set as* $t \to -\infty$ are defined in the same way.[7]

---

[4] In algebraic language, the set K is a group with respect to addition.

[5] It is easy to see that in this case, K consists of all integral multiples of T.

[6] In this case, K contains the whole t-axis.

[7] Sometimes limit points as $t \to -\infty$ and $t \to +\infty$ are called α-*limit points* and ω-*limit points*, respectively, and similarly for limit sets.

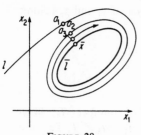

FIGURE 28

For example, suppose that as $t \to +\infty$ the trajectory l "spirals toward" a cycle $\bar{l}$ (see Figure 28). Then $\bar{l}$ is the limit set of l as $t \to +\infty$. In fact, choosing any point $\bar{x} \in \bar{l}$ and points

$$a_1 = x(t_1), \ldots, a_n = x(t_n), \ldots,$$

as shown in the figure, we find that

$$a_n = x(t_n) \to \bar{x} \qquad \text{as} \quad t_n \to \infty$$

(why?). A cycle like $\bar{l}$ which is the limit set as $t \to +\infty$ or as $t \to -\infty$ of another (distinct) trajectory is called a *limit cycle* (cf. Sec. 24).

It is easy to see that a critical point is its own unique limit point both as $t \to +\infty$ and as $t \to -\infty$. Moreover, a closed trajectory is its own limit set both as $t \to +\infty$ and as $t \to -\infty$. The limit sets of open trajectories are of greater interest, since they determine the behavior of trajectories for large $|t|$. We now establish some simple properties of limit sets, thinking of limit sets as $t \to +\infty$ (to be explicit).[8]

THEOREM 1. *A limit set $\bar{l}$ is closed, regarded as an n-dimensional point set, i.e., $\bar{l}$ contains all its (set-theoretic) limit points.*

Proof. Let $\bar{l}$ be the limit set of a trajectory l with equation $x = x(t)$, and consider a sequence $\bar{x}_k \in \bar{l}$, $\bar{x}_k \to \bar{x}$ as $k \to \infty$. By the definition of $\bar{l}$, given any k, there is a sequence of times $t_{k,n}$ such that $x(t_{k,n}) \to \bar{x}_k$ as $n \to \infty$. Let $\hat{t}_k$ be such that $\hat{t}_k > k$ and

$$\rho[x(\hat{t}_k), \bar{x}_k] < \frac{1}{k},$$

where $\rho(a, b)$ denotes the distance between two points a and b of x-space. Then clearly $\hat{t}_k \to \infty$ as $k \to \infty$, and moreover

$$\rho[x(\hat{t}_k), \bar{x}] \leqslant \rho[x(\hat{t}_k), \bar{x}_k] + \rho(\bar{x}_k, \bar{x}) < \frac{1}{k} + \rho(\bar{x}_k, \bar{x}) \to 0 \qquad \text{as} \quad k \to \infty.$$

In other words, $\bar{x}$ is a limit point of $x(t)$ as $t \to +\infty$, as required.

THEOREM 2. *If $\bar{x}$ belongs to a limit set $\bar{l}$, then $\bar{l}$ contains $l_{\bar{x}}$, i.e., $\bar{l}$ consists of whole "trajectories."*

Proof. Suppose the original trajectory l has equation $x = x(t; x^0)$, and let $x(t_n; x^0) \to \bar{x}$ as $n \to \infty$ (where $t_n \to \infty$). Then

$$x(t_n + t; x^0) = x[t; x(t_n; x^0)] \to x(t; \bar{x}),$$

i.e., $x(t; \bar{x}) \in \bar{l}$.

---

[8] Of course, limit sets as $t \to -\infty$ have just the same properties.

THEOREM 3. *A necessary and sufficient condition for a limit set to be empty is that*

$$\sum_{i=1}^{n} x_i^2(t) \to \infty \qquad as \quad t \to +\infty,$$

i.e., that the curve $x(t)$ "go off to infinity" as $t \to +\infty$.

Proof. If the condition holds, the limit set is obviously empty. Conversely, suppose the limit set is empty but the condition fails to hold. Then

$$\sum_{i=1}^{n} x_i^2(t) \leqslant R^2$$

for some R and sufficiently large t. Hence there is a sequence of times $t_k \to \infty$ $(k \to \infty)$ such that the sequence of points $x(t_k)$ is bounded. But a bounded sequence $x(t_k)$ contains a convergent subsequence, and the limit of this subsequence must be a limit point of $x(t)$ as $t \to +\infty$. This contradicts the assertion that the limit set is empty, thereby completing the proof.

THEOREM 4. *A necessary and sufficient condition for a limit set to consist of a single point $\bar{x}$ is that $x(t) \to \bar{x}$ as $t \to +\infty$, i.e., that the trajectory "enter the point $\bar{x}$" as $t \to +\infty$.*

Proof. If the condition holds, the limit point is obviously unique. Conversely, suppose the limit point is unique but the condition fails to hold. Then, given any $\varepsilon > 0$, there is a sequence $t_k \to +\infty$ $(k \to \infty)$ such that $\rho[x(t_k), \bar{x}] \geqslant \varepsilon$ for all t_k. But by the definition of $\bar{x}$, there is another sequence $t_k' \to +\infty$ such that $\rho[x(t_k'), \bar{x}] < \varepsilon$ for all sufficiently large t_k'. Therefore, by the continuity of $x(t)$, there must be a third sequence t_k'' such that $\rho[x(t_k''), \bar{x}] = \varepsilon$ for all sufficiently large t_k''. This bounded sequence t_k'' contains a subsequence converging to a point $\bar{\bar{x}}$ which is a limit point of $x(t)$ as $t \to +\infty$. But obviously $\rho(\bar{\bar{x}}, \bar{x}) = \varepsilon$, and hence there is at least one other limit point besides $\bar{x}$. This contradicts the assertion that $\bar{x}$ is the unique limit point, thereby completing the proof.

Problem. Prove that if a limit set $\bar{l}$ is nonempty and bounded, then it is connected, i.e., it cannot be represented as the union of two nonempty disjoint closed sets. (In particular, it follows that if $\bar{l}$ consists of more than one point, then it contains infinitely many points.) Show that if $\bar{l}$ is unbounded, then it may not be connected (give an example). However, show that $\bar{l}$ becomes connected if the "point at infinity in x-space" (i.e., the point "approached by every sequence of points x going off to infinity") is adjoined to $\bar{l}$.

54. Limiting Behavior of Trajectories: The Two-Dimensional Case

If $n = 2$, the system (7.2) reduces to

$$\frac{dx_1}{dt} = f_1(x_1, x_2), \qquad \frac{dx_2}{dt} = f_2(x_1, x_2). \tag{7.7}$$

Then, as shown by Bendixson, much more can be said about the limiting behavior of trajectories than in the case $n > 2$. We now give some of Bendixson's results, which are based on the fact that a closed curve in the plane (with no self-intersections) divides the plane into two parts.[9]

THEOREM 1.   *If a trajectory l contains at least one of its limit points $\bar{x}$ as $t \to +\infty$ or as $t \to -\infty$, then l is either a closed trajectory or a critical point.*

Proof.   To be explicit, let $t \to +\infty$. If $\bar{x}$ is a critical point, the whole trajectory reduces to the single point $\bar{x}$ (why?), and the theorem is proved.

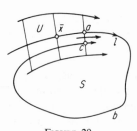

FIGURE 29

Thus suppose $\bar{x}$ is not a critical point, and consider a small neighborhood U of $\bar{x}$, bounded by two segments of normals to a small arc of the trajectory $l_{\bar{x}}$ containing $\bar{x}$ and by two arcs of neighboring trajectories (see Figure 29). The trajectories inside U form a family of slightly curved parallel arcs, along which the motion proceeds with almost constant velocity. By hypothesis, the trajectory l under discussion passes through $\bar{x}$. Since $\bar{x}$ is a limit point of l, the trajectory l must intersect U again as t is increased. We now show that l passes through $\bar{x}$ again as t is increased, i.e., that l is a closed trajectory. In fact, suppose l does not pass through $\bar{x}$ again. Then the arc of l together with the segment ca forms a closed curve (see Figure 29), bounding a finite region S of the plane, which we call a "Bendixson pocket." Suppose the trajectories of the system crossing ca enter S from the outside, as in Figure 29. Then l is doomed to stay inside S forever after, since it can neither intersect the arc abc (an open trajectory has no self-intersections) nor the segment ca (all trajectories crossing ca must enter S from the outside). But then after entering the pocket at c, the trajectory l can never come arbitrarily close to $\bar{x}$, contradicting the fact that $\bar{x}$ is a limit point. Similarly, if the trajectories crossing ca leave S from the inside, as in Figure 30, the

---

[9] This is not true in spaces of higher dimension.

trajectory l can never return to the pocket S after leaving it at c, and hence can never come arbitrarily close to $\bar{x}$, again contradicting the fact that $\bar{x}$ is a limit point of l. This completes the proof.

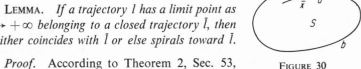

LEMMA. *If a trajectory l has a limit point as $t \to +\infty$ belonging to a closed trajectory $\bar{l}$, then l either coincides with $\bar{l}$ or else spirals toward $\bar{l}$.*

FIGURE 30

Proof. According to Theorem 2, Sec. 53, the limit set of l contains the whole trajectory $\bar{l}$. Consider a small neighborhood U of any point $\bar{x} \in \bar{l}$ (see Figure 31). By the definition of a limit point, the trajectory l must intersect U for arbitrarily large t, and hence, since the trajectories inside U are "almost parallel," l must intersect the normal $\bar{x}n$ to $\bar{l}$ (drawn through $\bar{x}$) at some point inside U. If the point of intersection is the point $\bar{x}$ itself, then $l = \bar{l}$.

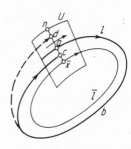

FIGURE 31

Otherwise, let the point of intersection be a point a, which, for definiteness, is assumed to lie outside $\bar{l}$. If a is close enough to $\bar{x}$, then the theorem on continuous dependence of a solution on the initial data implies that l will again fall in U after a time T equal to the period of traversing the cycle $\bar{l}$ once. Let c denote the first intersection of l with $\bar{x}n$ after a. Then the whole arc ac of l lies in an arbitrarily narrow strip along the cycle $\bar{l}$, provided that a is close enough to $\bar{x}$ (why?). It is easy to see that $\rho(c, \bar{x}) < \rho(a, \bar{x})$. In fact, if $c = a$, then l is a cycle and hence never approaches $\bar{x}$, while if l traverses the dashed path in Figure 31, it can never enter the "pocket" *abda* (why not?) and once again cannot approach $\bar{x}$. Continuing the trajectory l after c, we obtain another "loop" around $\bar{l}$ and another point of intersection of l with $\bar{x}n$, and so on. Since $\bar{x}$ is a limit point of l, these points of intersection must approach $\bar{x}$, and hence, because of the theorem on continuous dependence of a solution on the initial data, the consecutive loops of l must converge uniformly to $\bar{l}$. This proves the lemma.

THEOREM 2. *Let $\bar{G}$ be a closed bounded domain containing no critical points of the system (7.2), and suppose every trajectory beginning inside $\bar{G}$ at $t = 0$ stays inside $\bar{G}$ for all $t > 0$. Then every such trajectory is either a cycle, or else spirals toward a cycle as $t \to +\infty$. In particular, $\bar{G}$ must contain at least one cycle.*

Proof. Consider any *positive half-trajectory* l beginning inside $\bar{G}$, i.e., a curve $x = x(t; x^0)$ where $x^0 \in \bar{G}$ and $0 \leqslant t < \infty$. Since l is contained in $\bar{G}$, by hypothesis, it has at least one limit point $\bar{x} \in G$ as $t \to +\infty$, by Theorem 3 of Sec. 53. If $\bar{x} \in l$, then l is a closed trajectory, by Theorem 1 of this section. Otherwise, consider the positive half-trajectory $\bar{l}$ beginning at $\bar{x}$, and let $\bar{\bar{x}}$ be a limit point of $\bar{l}$ as $t \to +\infty$.

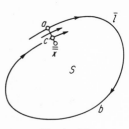

FIGURE 32

We need only show that $\bar{\bar{x}} \in \bar{l}$, since then Theorem 1 implies that $\bar{l}$ is a closed trajectory, and hence l spirals toward $\bar{l}$ by the lemma. If $\bar{\bar{x}}$ does not belong to $\bar{l}$, consider a small neighborhood of $\bar{\bar{x}}$ (see Figure 32). Let a and c be two points of intersection of $\bar{l}$ with the normal drawn through $\bar{\bar{x}}$ to $l_{\bar{\bar{x}}}$, and suppose the trajectories crossing the segment ac of the boundary $abca$ of the "pocket" S enter S from the outside, as in Figure 32. By Theorem 2 of Sec. 53, every point of $\bar{l}$ is a limit point of l. Therefore l eventually comes arbitrarily close to c and hence enters the pocket S. But then l must stay inside S forever after, and hence cannot approach the point $a \in \bar{l}$ (see Figure 32), despite the fact that a is a limit point of l as $t \to +\infty$. The case where the trajectories crossing ac leave S from the inside is treated similarly. This contradiction completes the proof.

THEOREM 3. *Given a closed trajectory l, suppose there are no other closed trajectories in some neighborhood of l. Then every trajectory beginning sufficiently close to l spirals toward l as $t \to +\infty$ or as $t \to -\infty$. Moreover, it is impossible for one trajectory outside l to spiral toward l as $t \to +\infty$ while another spirals toward l as $t \to -\infty$, and similarly for trajectories inside l.*

Proof. Since there are no critical points on l, there is a narrow strip D containing l in which there are also no critical points (why?), and moreover no cycles other than l. Let nn be a fixed, sufficiently small segment of the normal to l at some point $\bar{x} \in l$ (see Figure 33). All the trajectories beginning sufficiently near l come arbitrarily close to l when continued forwards and backwards in time, and hence intersect nn. Therefore we need only consider trajectories beginning on the segment nn near $\bar{x}$. Suppose we draw a trajectory through some point $a \in nn$. If b is the next point of intersection of this trajectory with nn, then

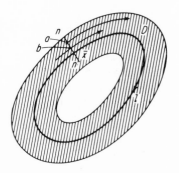

FIGURE 33

$b \neq a$ (why?). Assuming that $\rho(b, \bar{x}) < \rho(a, \bar{x})$, and considering the "ring-shaped pocket" bounded by the arc ab of the trajectory through a, the segment ba and the cycle l, we find that the next intersection of the trajectory with nn after b occurs at a point c on the segment $\bar{x}b$, the following one occurs on the segment $\bar{x}c$, and so on. The resulting sequence of points of intersection must converge to $\bar{x}$, since otherwise there would be a cycle distinct from l passing through the limit point of the sequence (think this through!). In other words, the given trajectory spirals toward l. The other trajectories beginning outside l (and sufficiently near l) are "squeezed between the loops" of the given trajectory, and hence also spiral toward l as $t \to +\infty$. On the other hand, if $\rho(b, \bar{x}) > \rho(a, \bar{x})$, we need only reverse the time direction (replacing t by $-t$) to reduce the problem to the case just considered, i.e., the trajectories now spiral toward l as $t \to -\infty$. Since the part of the strip D lying inside l can be handled in the same way, the proof is now complete.

Remark. It follows from Theorem 3 that an *isolated cycle*, i.e., a cycle which has a neighborhood containing no other cycles, must take one of the three forms shown in Figure 34, called *stable*, *unstable*, and *semistable*, respectively.

FIGURE 34

Problem 1. Give an example showing that Theorem 1 is false for an autonomous system in n-dimensional space if $n \geqslant 3$.

Problem 2. Give an example showing that a stable limit cycle need not be stable in the sense of Lyapunov. What extra conditions guarantee Lyapunov stability?

Problem 3. Suppose the right-hand sides of the autonomous system (7.7) are polynomials of degree no higher than two. First prove that no curve other than an integral curve can be tangent at more than two points to the direction field determined by (7.7). Using this fact, prove that

a) Every closed trajectory of (7.7) is convex;

b) The motion along two closed trajectories proceeds in opposite directions if one trajectory lies outside the other, and in the same direction if one lies inside the other.

Prove that there can be no more than one critical point inside a closed trajectory.

Comment. It can be shown, but not too easily, that a critical point lying inside a closed trajectory is a focus.[10]

55. The Succession Function

A useful concept in studying cycles and certain other problems is that of the succession function. In essence, this notion has already been used in Sec. 54. For simplicity, we confine our discussion to the case $n = 2$, corresponding to an autonomous system (7.7) in the plane. Suppose we are given a smooth curve L (see footnote 5, p. 6) which has no self-intersections and no "points of contact with trajectories of (7.7)," i.e., no points where L is tangent to a trajectory. For example, L might be a small segment of the normal to a trajectory, as in Sec. 54. Suppose the position of a variable point $a \in L$ is specified by the value of a parameter τ, so that $a = a(\tau)$.[11] Drawing the trajectory of (7.7) passing through the point $a(\tau_0)$ in the direction of increasing time, we continue the trajectory until its first intersection with L, if such an intersection occurs. This point of intersection corresponds to some parameter value $\tau = \tau_1$ depending on τ_0, and leads to a function $\tau_1 = \psi(\tau_0)$ called the *succession function*.

In general, $\psi(\tau)$ is not defined on the whole curve L, but only for values of τ such that the continuations of the corresponding trajectories intersect L again. In fact, $\psi(\tau)$ may not be defined at any point of L at all. However, if $\psi(\tau)$ is defined for a given τ_0, then it must be defined and continuous for all τ sufficiently near τ_0. This follows from the theorem on continuous dependence of a solution on the initial data, and from the fact that the curve L has no points of contact with trajectories (so that any trajectory meeting L must intersect L).

The following theorem shows how the succession function can be used to find cycles:

THEOREM. *A necessary and sufficient condition for a cycle to pass through the point $a(\tau_0)$ is that $\psi(\tau_0)$ be defined and that $\psi(\tau_0) = \tau_0$.*

Proof. The sufficiency of the condition is obvious. To prove the necessity, suppose first that $\psi(\tau_0)$ is not defined. Then the trajectory $l_{a(\tau_0)}$ never intersects L for $\tau > 0$ and in particular cannot pass through

---

[10] See Tung Chin-chu, *Positions of limit-cycles of the system*

$$\frac{dx}{dt} = \sum_{0 \leqslant i+k \leqslant 2} a_{ik}x^i y^k, \qquad \frac{dy}{dt} = \sum_{0 \leqslant i+k \leqslant 2} b_{ik}x^i y^k,$$

Scientia Sinica, **8**, 151 (1959).

[11] The letter t is reserved for the time, i.e., for the parameter of trajectories of the system.

$a(\tau_0)$ again. Therefore $l_{a(\tau_0)}$ cannot be a cycle in this case. Thus suppose that $\tau_1 = \psi(\tau_0)$ is defined but $\tau_1 \neq \tau_0$, and let $a(\tau_0) = a_0$, $a(\tau_1) = a_1$. Then the arc a_0a_1 of the trajectory l_{a_0} and the arc a_1a_0 of the curve L together form the boundary of a "Bendixson pocket," since L has no points of contact with trajectories and hence the trajectories crossing L all go in the same direction (why?). The arc a_1a_0 of the curve L can be intersected by trajectories entering the pocket as in Figure 35(a), or leaving it as in Figure 35(b). In either case, it is clear that the

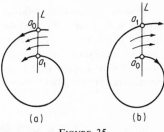

FIGURE 35

trajectory l_{a_0}, whose continuation passes through the point a_1, can never approach a_0 again. Therefore l_{a_0} cannot be a cycle, and the theorem is proved.

Remark. Suppose a cycle passes through the point $a(\tau_0)$, so that $\psi(\tau_0) = \tau_0$. Then the character of the stability of the cycle (recall Figure 34) depends on the behavior of $\psi(\tau)$ near the value $\tau = \tau_0$. If the cycle is isolated (i.e., if it has a neighborhood in which there are no other cycles), then $\psi(\tau) \neq \tau$ for sufficiently small $|\tau - \tau_0| > 0$. It is easily verified that the cycle is stable, unstable or semistable, depending on whether $\psi(\tau)$ has the behavior shown in Figure 36(a), 36(b) or 36(c).

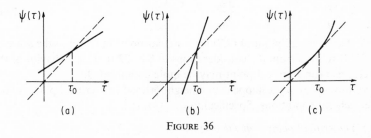

FIGURE 36

Problem 1. Prove that $\psi(\tau)$ is an increasing function.

Problem 2. Prove that if the system (7.7) has a stable limit cycle l, then, given any $\varepsilon > 0$, there exists a $\delta > 0$ such that any system obtained by changing the right-hand sides of (7.7) by less than δ has at least one cycle in an ε-neighborhood of l. Prove that if there are finitely many such cycles, then the number of stable cycles is one larger than the number of unstable cycles. How about the case where (7.7) has an unstable limit cycle? Prove that arbitrarily small changes in the right-hand sides of the system can lead to disappearance of a semistable cycle.

Problem 3. Examine the structure of a "sufficiently narrow" neighborhood of a nonisolated cycle. Prove that the neighborhood is completely filled by cycles or parts of cycles if the right-hand sides of the system are analytic.

Problem 4. Given an n-dimensional autonomous system, introduce the concept of a succession function for a smooth $(n-1)$-dimensional surface which has no points of contact with trajectories. Establish the relation between this function and cycles. Assuming that the mapping defining the succession function has a principal linear part in a neighborhood of a fixed point, give sufficient conditions for stability of the corresponding cycle.

56. Behavior Near a Critical Point

It is easy to see that a necessary and sufficient condition for a point $x = x^0$ to be a critical point of the autonomous system (7.2) is that

$$f(x^0) = 0 \tag{7.8}$$

[substitute $x(t) \equiv x^0$ into (7.2)]. Writing (7.8) in component form, we obtain n equations in the n unknown coordinates of the critical point. In the case of a two-dimensional autonomous system (7.7), the coordinates of the critical point satisfy the equations

$$f_1(x_1^0, x_2^0) = 0, \qquad f_2(x_1^0, x_2^0) = 0. \tag{7.9}$$

Since (7.7) and (7.9) together imply

$$\frac{dx_2}{dx_1} = \frac{f_2(x_1, x_2)}{f_1(x_1, x_2)}, \tag{7.10}$$

we see that a critical point of (7.7) is a singular point of (7.10) in the sense of Sec. 22. It follows from the considerations of Sec. 22 that the behavior of the trajectories near a critical point may be quite complicated.

We now study what happens in a neighborhood of a critical point, under rather general conditions. Specifically, we assume that

1) *The critical point is at the origin;*[12]
2) *Each of the functions f_1, f_2 appearing in (7.7) can be represented as the sum of a homogeneous polynomial of degree $m \geqslant 1$ and higher-order terms, i.e.,*

$$f_i(x_1, x_2) = P_i(x_1, x_2) + \psi_i(x_1, x_2), \qquad i = 1, 2, \tag{7.11}$$

where[13]

$$\psi_i(x_1, x_2) = o(|x_1|^m + |x_2|^m).$$

---

[12] This can always be achieved by parallel displacement of the coordinate axes.
[13] For the meaning of the symbol o, see Prob. 6, Sec. 49.

It is convenient to first transform to polar coordinates ρ and φ, related to x_1 and x_2 by the formulas

$$x_1 = \rho \cos \varphi, \qquad x_2 = \rho \sin \varphi.$$

Then the system (7.7) becomes

$$\frac{d\rho}{dt} = \frac{1}{\rho}\left(x_1 \frac{dx_1}{dt} + x_2 \frac{dx_2}{dt}\right) = \frac{1}{\rho}(x_1 P_1 + x_2 P_2) + o(\rho^m),$$

$$\frac{d\rho}{dt} = \frac{1}{\rho^2}\left(x_1 \frac{dx_2}{dt} - x_2 \frac{dx_1}{dt}\right) = \frac{1}{\rho^2}(x_1 P_2 - x_2 P_1) + o(\rho^{m-1}),$$

or

$$\frac{d\rho}{dt} = \rho^m Q(\varphi) + o(\rho^m), \qquad \frac{d\varphi}{dt} = \rho^{m-1}R(\varphi) + o(\rho^{m-1}) \qquad (7.12)$$

in terms of the functions

$$Q(\varphi) = P_1(\cos \varphi, \sin \varphi) \cos \varphi + P_2(\cos \varphi, \sin \varphi) \sin \varphi,$$

$$R(\varphi) = P_2(\cos \varphi, \sin \varphi) \cos \varphi - P_1(\cos \varphi, \sin \varphi) \sin \varphi.$$

These functions are both homogeneous polynomials in $\cos \varphi$ and $\sin \varphi$ (of degree $m + 1$), and hence are periodic with period 2π.

1. *First we consider the case where $R(\varphi) \neq 0$ $(0 \leqslant \varphi < 2\pi)$, say $R(\varphi) > 0$.* Then, according to the second of the equations (7.12), in a small neighborhood of the origin every trajectory moves around the origin in the positive direction as t increases. Moreover, it follows from (7.12) that

$$\frac{1}{\rho}\frac{d\rho}{d\varphi} = \frac{Q(\varphi) + o(1)}{R(\varphi) + o(1)}, \qquad (7.13)$$

and hence, integrating along any trajectory, we find that

$$\ln \rho(\alpha + 2\pi) - \ln \rho(\alpha) = \int_\alpha^{\alpha+2\pi} \frac{Q(\varphi) + o(1)}{R(\varphi) + o(1)} \, d\varphi. \qquad (7.14)$$

Suppose first that

$$I = \int_0^{2\pi} \frac{Q(\varphi)}{R(\varphi)} \, d\varphi < 0. \qquad (7.15)$$

Integrating (7.13) between the limits α and $\alpha + \varphi$ $(0 \leqslant \varphi \leqslant 2\pi)$, instead of between α and $\alpha + 2\pi$, we find that the difference $\ln \rho(\alpha + \varphi) - \ln \rho(\alpha)$ is bounded in a small neighborhood U of the origin, and hence the ratio

$$\frac{\rho(\alpha + \varphi)}{\rho(\alpha)} \qquad (0 \leqslant \varphi \leqslant 2\pi)$$

is bounded both from above and from below by positive constants. Thus if a trajectory begins sufficiently near the origin, it remains in U after making a complete circuit around the origin. But if the neighborhood U is small enough, the right-hand side of (7.14) will be arbitrarily close to I, say between $\frac{1}{2}I$ and $\frac{3}{2}I$. Then (7.15) implies that $\ln \rho \to -\infty$ and hence $\rho \to 0$ as $\varphi \to +\infty$. Hence, under these conditions, every trajectory beginning sufficiently near the critical point spirals toward the point as $t \to +\infty$, winding around the point an infinite number of times in the positive direction. In this case, the critical point is called a *stable focus* (cf. Sec. 22).

Similarly, if $I > 0$, it can be shown that the trajectories all spiral toward the critical point as $t \to -\infty$, i.e., the critical point is an *unstable focus*. If $I = 0$, it can only be shown (do so!) that if any trajectory l approaches the critical point as $t \to +\infty$ or as $t \to -\infty$, then l is a spiral trajectory and moreover so is every other trajectory beginning near the critical point (which is again a focus). However, if $I = 0$ it may happen that every neighborhood of the critical point contains closed trajectories surrounding the critical point. Then if all the trajectories are closed for sufficiently small $\rho > 0$, the critical point is a *center* (cf. Sec. 22). Otherwise the critical point is called a "center-focus," but Poincaré has shown that a critical point of this kind is impossible if the functions $f_1(x_1, x_2)$ and $f_2(x_1, x_2)$ are analytic. Which case actually occurs, a center or a focus, can only be ascertained by studying the higher-order terms ψ_i in (7.11). This complicated problem will not be gone into here.

2. *Next we consider the case where* $R(\varphi)$ *takes values of both signs.* Then $R(\varphi)$ has no more than $m + 1$ zeros for $0 \leqslant \varphi < \pi$ (why?), and shifting these zeros by π gives the zeros of $R(\varphi)$ for $\pi \leqslant \varphi < 2\pi$. In this case, every trajectory approaching the critical point as $t \to +\infty$ or as $t \to -\infty$ has a definite direction $\tilde{\varphi}$ at the critical point, where $R(\tilde{\varphi}) = 0$, and hence there can be no more than $2m + 2$ directions of approach to the critical point.

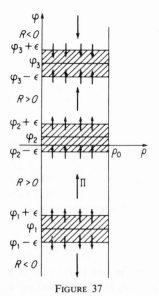

FIGURE 37

To prove these assertions, we represent the trajectories in an auxiliary $\rho\varphi$-plane (see Figure 37). If a trajectory l approaches the critical point, then, starting from a certain time, l must lie in the strip $0 < \rho \leqslant \rho_0$, where ρ_0 is any preassigned positive number. Given any $\varepsilon > 0$, we draw the lines $\varphi = \varphi_k \pm \varepsilon$, where the φ_k denote the zeros of $R(\varphi)$. If ρ_0 is small enough, the trajectory can intersect each of the unshaded rectangles shown in the figure in only one direction, either upward or downward (why?). On the other hand, integrating (7.13) from φ^* to $\tilde{\varphi}$, where $\varphi_1 + \varepsilon \leqslant \varphi^* < \tilde{\varphi} \leqslant \varphi_2 - \varepsilon$, we get an upper bound for the factor by which ρ changes as the point (ρ, φ) traverses the unshaded rectangle Π. Then the second of the equations (7.12) implies that once having entered the rectangle Π, a trajectory must leave Π within finite time, crossing the line $\varphi = \varphi_2 - \varepsilon$, and similarly for every other unshaded rectangle. But as $\rho \to 0$, a trajectory can intersect only a finite number of these rectangles (why?). Therefore, starting from some sufficiently small value of ρ, a trajectory approaching the critical point must stay forever in one of the shaded rectangles. Since ε can be made arbitrarily small, this proves that φ approaches one of the φ_k as $\rho \to 0$.

3. *Now let* $R(\bar{\varphi}) = 0$, $Q(\bar{\varphi}) \neq 0$, *and suppose* $R(\varphi)$ *changes sign in passing through* $\bar{\varphi}$. In this case, at least one trajectory approaches the critical point along the direction $\bar{\varphi}$. Moreover, let S be an angular sector with vertex at the critical point and bisector $\varphi = \bar{\varphi}$. Then every trajectory in S approaches the critical point along the direction $\bar{\varphi}$, if the radius and angle of S are sufficiently small and if $R(\varphi)Q(\varphi)$ changes sign from $-$ to $+$ in passing through $\bar{\varphi}$.

The proof of these assertions goes as follows: It can be assumed that $Q(\bar{\varphi}) < 0$, since otherwise we need only replace t by $-t$, which causes R and Q to go into $-R$ and $-Q$ without affecting the sign of RQ. Suppose $R(\varphi)$ changes sign from $+$ to $-$. Then for sufficiently small $\varepsilon > 0$ and $\rho_0 > 0$, the velocity field on the boundary of the rectangle $\tilde{\Pi}: |\varphi - \bar{\varphi}| \leqslant \varepsilon, 0 \leqslant \rho \leqslant \rho_0$ (see Figure

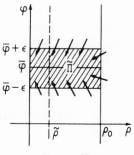

FIGURE 38

38) points into $\tilde{\Pi}$ if $\rho > 0$ ($\tilde{\Pi}$ corresponds to a sector in the $x_1 x_2$-plane). This means that the trajectories beginning in $\tilde{\Pi}$ cannot leave $\tilde{\Pi}$ as t increases. But it follows from the first of the equations (7.12) that every such trajectory reaches any line $\rho = \tilde{\rho}$ (if $\tilde{\rho} > 0$ is sufficiently small) in finite time, and hence approaches the critical point as $t \to +\infty$, where, as shown above, the direction of approach is $\bar{\varphi}$.

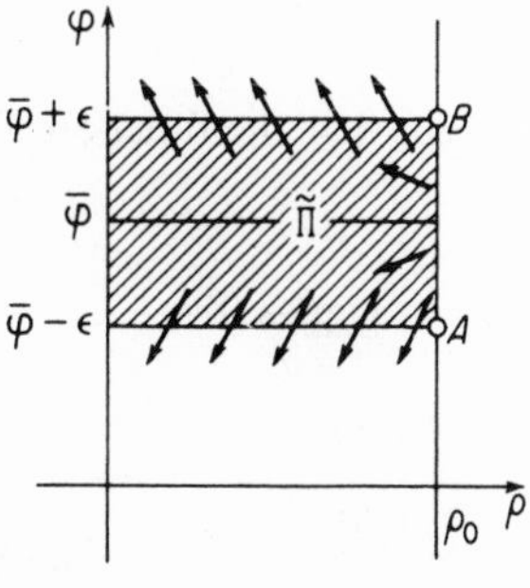

On the other hand, if $R(\varphi)$ changes sign from $-$ to $+$, then, for $\rho > 0$, the velocity field on the boundary of the rectangle $\tilde{\Pi}$ is directed as shown in Figure 39. Consider the trajectories beginning on the segment AB. According to the theorem on continuous dependence of the solution on the initial data, if one of these trajectories leaves $\tilde{\Pi}$ through the upper (or lower) side, then so does every other

FIGURE 39

trajectory beginning sufficiently near the given trajectory. Let C be the least upper bound of all points of AB at which trajectories begin and subsequently leave $\widetilde{\Pi}$ through the lower side. Then the trajectory beginning at C can never leave $\widetilde{\Pi}$ (why not?). Therefore, by the previous argument, this trajectory approaches the critical point along the direction $\bar{\varphi}$, and the proof of our assertions is now complete.

4. *Again let* $R(\bar{\varphi}) = 0$, $Q(\bar{\varphi}) \neq 0$, *and suppose* $R(\varphi)$ *changes sign from* $-$ *to* $+$ *in passing through* $\bar{\varphi}$. *Suppose further that* $R'(\bar{\varphi}) \neq 0$ *and that the functions* ψ_i *appearing in the system* (7.11) *have continuous first derivatives with respect to* x_1 *and* x_2 *of order* $o(\rho^{m-1})$ *as* $\rho \to 0$. *Then only one trajectory approaches the critical point along the direction* $\bar{\varphi}$. To see this, we first write (7.13) in more detail as

$$\rho \frac{d\varphi}{d\rho} = \frac{\rho^{m+1}R(\varphi) + x_1\psi_2 - x_2\psi_1}{\rho^{m+1}Q(\varphi) + x_1\psi_1 + x_2\psi_2} \equiv U(\rho, \varphi). \tag{7.16}$$

A direct calculation shows that the right-hand side of (7.16) has a continuous derivative with respect to φ in $\widetilde{\Pi}$ (see Figure 38) if $\rho > 0$, where this derivative can be made arbitrarily close to

$$\frac{R'(\bar{\varphi})}{Q(\bar{\varphi})} \tag{7.17}$$

if ε and ρ_0 are sufficiently small. Now suppose two distinct trajectories with equations

$$\varphi = \varphi_1(\rho), \qquad \varphi = \varphi_2(\rho) \tag{7.18}$$

approach the critical point along the direction $\bar{\varphi}$. Substituting (7.18) into (7.16) and subtracting the first of the resulting equations from the second, we obtain

$$\rho \frac{d(\varphi_2 - \varphi_1)}{d\rho} = U(\rho, \varphi_2) - U(\rho, \varphi_1). \tag{7.19}$$

According to Hadamard's lemma, the right-hand side of (7.19) can be written in the form

$$(\varphi_2 - \varphi_1)F(\rho, \varphi_1, \varphi_2),$$

where it clear from the proof of the lemma (see Sec. 20) that the function F is continuous in all its arguments on $\widetilde{\Pi}$ if $\rho > 0$ and

$$F(\rho, \varphi_1, \varphi_2) \to \frac{R'(\bar{\varphi})}{Q(\bar{\varphi})} < 0 \qquad \text{as} \quad \varphi_1 \to \bar{\varphi}, \varphi_2 \to \bar{\varphi}, \rho \to 0+.$$

Thus (7.19) becomes

$$\rho \frac{d(\varphi_2 - \varphi_1)}{d\rho} = (\varphi_2 - \varphi_1)F[\rho, \varphi_1(\rho), \varphi_2(\rho)]$$
$$\equiv A(\rho)(\varphi_2 - \varphi_1) \qquad (0 < \rho \leqslant \rho_0), \tag{7.20}$$

where the function $A(\rho)$ is continuous and approaches the limit (7.17) as $\rho \to 0+$. Integrating (7.20) as a homogeneous linear equation (see Sec. 7), we find that

$$\varphi_2 - \varphi_1 = C \exp \left\{ \int_{\rho_0}^{\rho} \frac{A(r)}{r} \, dr \right\} = C \exp \left\{ \int_{\rho_0}^{\rho} \frac{A(r)}{r} \, dr \right\}.$$

But the last integral approaches $+\infty$ as $\rho \to 0+$ (why?), and hence C must vanish if the left-hand side is to remain finite. In other words $\varphi_1(\rho) \equiv \varphi_2(\rho)$, contrary to the assumption that the two solutions (7.18) are distinct. Therefore as asserted, only one trajectory approaches the critical point along the direction $\bar{\varphi}$.

5. *Finally let $R(\varphi) \equiv 0$, $Q(\varphi) \not\equiv 0$, and suppose that each of the functions ψ_i appearing in the system* (7.11) *can be represented as the sum of a homogeneous polynomial of degree $m + 1$ and higher-order terms.* Then at least one trajectory approaches the critical point along every direction $\bar{\varphi}$ for which $Q(\bar{\varphi}) \neq 0$ [there are no more than $2m + 2$ directions for which $Q(\bar{\varphi}) = 0$]. In fact, under these conditions, dividing both sides of (7.16) by ρ leads to an expression on the right which is continuous everywhere in $\tilde{\Pi}$, including the left side of $\tilde{\Pi}$. Then our assertion follows from the theorem on the existence of a solution of a differential equation with a continuous right-hand side, satisfying the initial conditions $\rho = 0$, $\varphi = \bar{\varphi}$.[14]

Example. We now apply these results to the most important case $m = 1$. Then the system (7.7) takes the form

$$\frac{dx_1}{dt} = ax_1 + bx_2 + o(\rho), \qquad \frac{dx_2}{dt} = cx_1 + dx_2 + o(\rho), \qquad (7.21)$$

and the functions $Q(\varphi)$, $R(\varphi)$ are the following homogeneous polynomials in $\cos \varphi$ and $\sin \varphi$:

$$\begin{aligned} Q(\varphi) &= (a \cos \varphi + b \sin \varphi) \cos \varphi + (c \cos \varphi + d \sin \varphi) \sin \varphi \\ &= a \cos^2 \varphi + (b + c) \cos \varphi \sin \varphi + d \sin^2 \varphi, \qquad (7.22) \\ R(\varphi) &= c \cos^2 \varphi + (d - a) \cos \varphi \sin \varphi - b \sin^2 \varphi. \end{aligned}$$

We shall assume that

$$ad - bc \neq 0, \qquad (7.23)$$

since otherwise $Q(\varphi)$ and $R(\varphi)$ have common zeros,[15] a case which was not considered above. Note that the condition (7.23) was also imposed in Sec. 22.

---

[14] Suppose it is also known that the functions ψ_i have continuous partial derivatives with respect to x_1 and x_2 of order $O(\rho^m)$ as $\rho \to 0$. Then it is not hard to see that the derivative of the right-hand side of (7.16) is of order $O(\rho)$, and hence only one integral curve can approach the critical point along the direction $\bar{\varphi}$.

[15] In fact, $Q(\varphi)$ and $R(\varphi)$ have the common factor $c \cos \varphi + d \sin \varphi$ if $|c| + |d| > 0$, and the common factor $a \cos \varphi + b \sin \varphi$ if $c = d = 0$

If

$$(d - a)^2 + 4bc < 0,$$

then a direct calculation shows that

$$\int_0^{2\pi} \frac{Q(\varphi)}{R(\varphi)} \, d\varphi = -2\pi \frac{b}{|b|} \frac{a + d}{\sqrt{-(d - a)^2 - 4bc}}$$

(verify this). Therefore the origin is a focus if $a + d \neq 0$, and the integral curves wind around the origin in the positive direction if $b(a + d) > 0$ and in the negative direction if $b(a + d) < 0$. Since $d\varphi/dt$ and b have opposite signs, we see that the focus is stable if $a + d > 0$ and unstable if $a + d < 0$ (deduce this result from the theorem of Sec. 49). However, if $a + d = 0$, we can have a center, a focus or a center-focus, depending on the behavior of the nonlinear terms in the system (7.21).

Next let

$$(d - a)^2 + 4bc > 0 \tag{7.24}$$

Then the function $R(\varphi)$ with period π given by (7.22) has two zeros in the interval $0 \leqslant \varphi < \pi$, and changes sign passing through each of these zeros. Therefore the zeros of $R(\varphi)$ in the interval $0 \leqslant \varphi < 2\pi$ determine two pairs of opposite directions, along each of which at least one integral curve approaches the origin as $t \to +\infty$ or as $t \to -\infty$. To study this case in more detail, we note that (7.24) implies that the characteristic equation

$$\begin{vmatrix} a - \lambda & b \\ c & d - \lambda \end{vmatrix} = 0,$$

i.e.,

$$\lambda^2 - (a + d)\lambda - ad - bc = 0, \tag{7.25}$$

has distinct real roots. Therefore, according to Sec. 48, there is a nonsingular linear transformation with real coefficients reducing the system (7.21) to the form

$$\frac{d\xi_1}{dt} = \lambda_1 \xi_1 + o(\rho^*), \qquad \frac{d\xi_2}{dt} = \lambda_2 \xi_2 + o(\rho^*), \tag{7.26}$$

where λ_1 and λ_2 are the roots of equation (7.25), and $\rho^* = \sqrt{\xi_1^2 + \xi_2^2}$ is the radial distance in the new $\xi_1 \xi_2$-plane. Instead of (7.22), we now have

$$Q^*(\varphi^*) = \lambda_1 \cos^2 \varphi^* + \lambda_2 \sin^2 \varphi^*,$$
$$R^*(\varphi^*) = (\lambda_2 - \lambda_1) \cos \varphi^* \sin \varphi^*,$$

where φ^* is the polar angle in the $\xi_1 \xi_2$-plane. It follows from the second of these equations that the integral curves approach the origin along the coordinate axes. Here we must distinguish two subcases. If $\lambda_1 \lambda_2 > 0$, i.e.,

$$ad - bc > 0, \tag{7.27}$$

and if $\lambda_1 + \lambda_2 = a + d < 0$, then domains of the kind shown in Figures 38 and 39 alternate in the $\rho^*\varphi^*$-plane, and hence the origin is a stable node (i.e., the motion along the trajectories is directed toward the critical point at the origin). If, in addition, the right-hand sides of the system (7.21) are continuously differentiable, then the conditions in italics on p. 180 are satisfied (why?), and hence one integral curve approaches the origin along each of two opposite directions while every other integral curve sufficiently near the origin approaches the origin along one of the other two opposite directions (see Figure 12, p. 68). On the other hand, if (7.27) holds but $a + d > 0$, then the origin is an unstable node. The other subcase $ad - bc < 0$ is treated in Problem 1 below.

Finally suppose that

$$(d - a)^2 + 4bc = 0.$$

Then there are again two subcases. If $|b| + |c| > 0$, the function $R(\varphi) \not\equiv 0$ does not change sign but has zeros. Although this case has not yet been considered, it follows from Problem 3 below that the origin is a node with two directions of approach if

$$\psi_i = O(\rho^{1+\varepsilon}), \qquad i = 1, 2$$

for some $\varepsilon > 0$ (see Figure 15, p. 69). On the other hand, if $b = c = 0$, then

$$a = d, \quad R(\varphi) \equiv 0, \quad Q(\varphi) \equiv a \neq 0.$$

It follows from our previous considerations that if each of the functions ψ_i can be represented as the sum of a homogeneous polynomial of degree 2 and higher-order terms, then the origin is a node which is approached by integral curves along every direction.

Problem 1. Prove that if $ad - bc < 0$ and if the right-hand sides of the system (7.21) are continuously differentiable, then the origin is a saddle point (see Figure 14, p. 69).

Problem 2. Give an example of an autonomous system with a critical point of the center-focus type. Examine the structure of a neighborhood of a critical point of this type in the general case.

Problem 3. Analyze the case where $R(\varphi) \not\equiv 0$ does not change sign but has zeros. In particular, assuming that $\psi_i = O(\rho^{m+\varepsilon})$, $i = 1, 2$ for some $\varepsilon > 0$, $R(\bar{\varphi}) = 0$, $Q(\bar{\varphi}) \neq 0$ and examining the direction of the velocity field on the lines $\varphi = \bar{\varphi} \pm \rho^h$ where $0 < h < \varepsilon$, prove that infinitely many integral curves approach the origin along the direction $\varphi = \bar{\varphi}$ as $t \to +\infty$ or as $t \to -\infty$. More exactly, if for definiteness $R(\varphi) \geqslant 0$, $Q(\bar{\varphi}) < 0$ and if $\tilde{\varphi}$ is the first zero of $R(\varphi)$ for $\varphi < \bar{\varphi}$, then for any φ^*, $\tilde{\varphi} < \varphi^* < \bar{\varphi}$, there exists a $\rho^* > 0$ such that every integral curve beginning at $0 < \rho < \rho^*$, $\varphi = \varphi^*$ approaches the origin along the direction $\varphi = \bar{\varphi}$ as $t \to +\infty$.

Problem 4. Classify the isolated critical points of an autonomous system for the case $n = 3$, assuming that the right-hand sides are linear and that the characteristic equation has distinct roots. Examine each of the five kinds of critical points from a geometric point of view. Carry out an analogous classification for arbitrary n.[16] Which of the critical points are "crude," i.e., do not change their type if the coefficients of the system are changed only slightly?

57. Theory of Indices

DEFINITION 1. *Let $f(x) = f(x_1, x_2)$ be a continuous vector field in the plane, with components $f_1(x_1, x_2)$, $f_2(x_1, x_2)$, and let L be a piecewise smooth oriented Jordan curve[17] passing through no critical points of $f(x)$.[18] Moreover, let $\theta = \theta(x)$ be the angle which $f(x)$ makes with a fixed direction l, and let $\Delta\theta$ be the change in θ as the point $A = (x_1, x_2)$ traverses L once in the positive direction. Then the quantity*

$$I_f(L) \equiv \frac{\Delta\theta}{2\pi}$$

is called the index of L with respect to f.

Remark 1. The concept of index is due to Poincaré and can be used to study two-dimensional autonomous systems, in which case the components of $f(x)$ are the right-hand sides of the system (7.7). However, the considerations that follow apply equally well to any continuous vector field in the plane.[19]

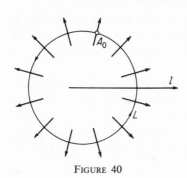

FIGURE 40

Remark 2. In calculating $\Delta\theta$, it is assumed that we choose a definite branch of the multiple-valued function $\theta = \theta(x)$. Thus the angle is a continuous function but there is no unique choice of the initial value of θ. For example, in the case of the vector field and curve shown in Figure 40, $\Delta\theta = 2\pi$ and hence $I_f(L) = 1$, regardless of whether the value of the angle at the initial point A_0 is taken to be $5\pi/12$, $29\pi/12$, $-19\pi/12$, etc.

---

[16] For a discussion of critical points of nonlinear autonomous systems for arbitrary n see V. V. Nemytski and V. V. Stepanov, *op. cit.*, Chap. 4.

[17] L is said to be *Jordan* if it has no self-intersections, and *oriented* if a definite direction along L is regarded as positive. We assume that L contains its end points, which coincide if L is closed. We shall always take the positive direction along a closed curve L to be the counterclockwise direction.

[18] By a *critical point* of $f(x)$, we mean a point where $f_1(x) = f_2(x) = 0$, just as on p. 176.

[19] Concerning this whole subject, see M. A. Krasnoselski, A. I. Perov, A. I. Povolotski and P. P. Zabreiko, *Vector Fields in the Plane* (in Russian), Gos. Izd. Fiz.-Mat. Lit., Moscow (1963).

Remark 3. It is easy to see that $I_f(L)$ does not depend on the direction l, nor on the initial point A_0 if L is closed.

Next we list some simple properties of $I_f(L)$:

1. Obviously, if $-L$ denotes L traversed in the opposite direction, then

$$I_f(-L) = -I_f(L). \tag{7.28}$$

Moreover, if $L = L_1 + \cdots + L_n$,[20] then

$$I_f(L) = I_f(L_1) + \cdots + I_f(L_n). \tag{7.29}$$

2. If f_1 and f_2 are continuously differentiable, then

$$I_f(L) = \frac{1}{2\pi} \int_L d\theta = \frac{1}{2\pi} \int_L d\left(\arctan \frac{f_2}{f_1}\right) = \frac{1}{2\pi} \int_L \frac{f_1 \, df_2 - f_2 \, df_1}{f_1^2 + f_2^2}$$

$$= \frac{1}{2\pi} \int_L \frac{1}{f_1^2 + f_2^2} \left[\left(f_1 \frac{\partial f_2}{\partial x_1} - f_2 \frac{\partial f_1}{\partial x_1}\right) dx_1 + \left(f_1 \frac{\partial f_2}{\partial x_2} - f_2 \frac{\partial f_1}{\partial x_2}\right) dx_2\right].$$

3. If L is closed, then $I_f(L)$ is an integer. In fact, since the initial point and final point of L coincide, the vector field $f(x)$ returns to its original value after traversing L once, i.e., $\Delta\theta = 2m\pi$ where m is an integer.

4. If L is closed and if L is continuously deformed into another closed curve L' without ever passing through a critical point of $f(x)$, then

$$I_f(L') = I_f(L).$$

In fact, under such a deformation, both L and $f(x)$ vary continuously, and hence so does $I_f(L)$. But then $I_f(L)$ cannot change at all, since, as just shown, $I_f(L) = 2m\pi$.

5. If L is closed and if there are no critical points of $f(x)$ inside L, then $I_f(L) = 0$. To see this, we deform L into an arbitrarily small circle L' inside L, with center a. Then, according to property 4, $I_f(L) = I_f(L')$. But $I_f(L') = 0$, since the values of $f(x)$ on L' are uniformly close to its value at a, and $f(a) \neq 0$ [think this through].

THEOREM 1. *If L is a closed trajectory of the two-dimensional autonomous system* (7.7), *then $I_f(L) = 1$.*

Proof. In this case, the components of $f(x)$ are given by the right-hand sides of (7.7). The curve L is a smooth Jordan curve (why?), and moreover $f(x)$ is tangent to L and vanishes nowhere on L. Whether or not the positive direction along L corresponds to the direction of

---

[20] In writing $L_1 + \cdots + L_n$, it is assumed that the final point of L_k ($k = 1, \ldots, n - 1$) coincides with the initial point of L_{k+1}.

increasing t [and hence to the direction of $f(x)$ itself], $f(x)$ makes one turn in the counterclockwise direction as L is traversed in the positive direction (recall footnote 17).[21] Therefore $I_f(L) = 1$, as asserted.

COROLLARY.   *Every closed trajectory of the system* (7.7) *contains at least one critical point of the system.*

Proof.   Use property 5.

DEFINITION 2.   *Let x^0 be critical point of $f(x)$, and let L be any closed Jordan curve containing x^0 and no other critical points.*[22] *Then the quantity*

$$I_f(x^0) \equiv I_f(L)$$

is called the index of x^0 with respect to f.[23]

THEOREM 2.   *If L is a closed Jordan curve passing through no critical points of the system* (7.7), *and if there are only finitely many critical points $x^1, \ldots, x^n$ inside L, then*

$$I_f(L) = \sum_{k=1}^{n} I_f(x^k). \tag{7.30}$$

Proof.   Draw extra curves inside L dividing the interior of L into subdomains, each containing only one critical point (see Figure 41). This gives n closed Jordan curves $L_1, \ldots, L_n$, where, as always, the positive direction of traversing each L_k is the counterclockwise direction. It follows from (7.28) and (7.29) that

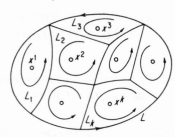

FIGURE 41

$$I_f(L) = \sum_{k=1}^{n} I_f(L_k)$$

(why?). But this, together with Definition 2, implies (7.30).

Problem 1.   Let $P(z)$ be a polynomial in $z = x + iy$ of degree at least 1. Writing $P = Q + iR$ and examining $I_f(L)$, where f is the vector field (Q, R) and L is a sufficiently large circle with center at the origin, prove that $P(z)$ has at least one zero.

Problem 2.   Prove that a node, a center or a focus has index 1, while a saddle point has index -1. Prove that if L is a closed trajectory containing nodes,

---

[21] For a rigorous proof of this intuitive geometric fact, see e.g., S. Lefschetz, *op. cit.*, p. 199.
[22] This presupposes that x^0 is an isolated point.
[23] In making this definition, we tacitly rely upon property 4.

centers, foci and saddle points, then the total number of critical points inside L is odd and the number of saddle points is one less than the total number of all other kinds of critical points.

Problem 3. It can be proved that an isolated critical point x^0, which is not a center or a focus, has a sufficiently small neighborhood $\mathcal{N}$ with the following structure: $\mathcal{N}$ contains a finite number of "elliptic sectors" completely filled by trajectories with x^0 at both ends, and a finite number of "hyperbolic sectors" completely filled by trajectories with the boundary of $\mathcal{N}$ at both ends. Both kinds of domains extend from x^0 to the boundary of $\mathcal{N}$ and are separated from each other by "fans" completely filled by trajectories with one end at x^0 and the other end on the boundary of $\mathcal{N}$. If the number of domains of each kind is known, what is the index of the critical point?

Problem 4. The concept of the index of a curve with respect to a vector field can be extended to the case of n dimensions. Let S be a closed piecewise smooth oriented[24] $(n - 1)$-dimensional surface, or a piece of such a surface satisfying similar requirements. Let

$$f(x) = (f_1(x), \ldots, f_n(x))$$

be a continuously differentiable vector function defined in a neighborhood of S. Then by the index of S with respect to f is meant the quantity

$$I_f(S) = \frac{1}{\omega_{n-1}} \int_S \sum_{i=1}^{n} \begin{vmatrix} \dfrac{\partial f_1}{\partial x_1} \cdots \dfrac{\partial f_1}{\partial x_{i-1}} & f_1 & \dfrac{\partial f_1}{\partial x_{i+1}} \cdots \dfrac{\partial f_1}{\partial x_n} \\ \cdots\cdots\cdots\cdots & \cdots & \cdots\cdots\cdots\cdots \\ \dfrac{\partial f_n}{\partial x_1} \cdots \dfrac{\partial f_n}{\partial x_{i-1}} & f_n & \dfrac{\partial f_n}{\partial x_{i+1}} \cdots \dfrac{\partial f_n}{\partial x_n} \end{vmatrix} \cos(\nu, x_i) \left(\sum_{i=1}^{n} f_i^2 \right)^{-n/2} dS,$$

where ω_{n-1} is the "volume" of the unit $(n - 1)$-dimensional sphere, and ν is the exterior normal to S. Give examples illustrating the concept of index in three dimensions, making the appropriate geometric interpretation. Prove some properties of $I_f(S)$.

Problem 5. Define the index of an isolated critical point of an n-dimensional autonomous system. Prove that the index equals ± 1 if the right-hand sides of the system are linear (what determines the sign?). What are the implications for the index of a critical point for a nonlinear system with a linear principal part?

Problem 6. Consider a continuously differentiable field of tangent vectors on a three-dimensional sphere R, and a corresponding autonomous system on R. Define the concept of the index of a critical point, and prove that if there are finitely many critical points, then the sum of their indices equals 2. Use this and a suitable approximation to prove that every continuous field of tangent vectors on R has at least one null vector. Also study the case where R is the surface of a torus or a "pretzel" (a sphere with two handles).

---

[24] I.e., with the outside indicated.

58. Brouwer's Fixed Point Theorem

We begin with a few general considerations. Two n-dimensional point sets K and K' are said to be *homeomorphic* (to each other) if there exists a continuous mapping

$$x' = \varphi(x) \in K' \qquad (x \in K)$$

of K onto K' such that the inverse mapping $x = \varphi^{-1}(x')$ maps K' continuously onto K.[25] For example, the n-dimensional ball (solid sphere) is homeomorphic to the n-dimensional cube or to the n-dimensional tetrahedron (or even to any convex n-dimensional body), but not to the set consisting of two disjoint balls. The properties common to all homeomorphic sets are studied in a special branch of mathematics called *topology*. From the topological point of view, homeomorphic sets are equivalent, just as congruent figures are regarded as equivalent in elementary geometry. Thus such concepts as angle, length, etc. are not topological. On the other hand, such "cruder" properties as connectivity, dimension, etc., are topological (although the proofs are often far from easy).

In proving the main result of this section, we shall need the following

LEMMA (*Sperner's lemma*).   *Suppose a given triangle ABC is divided into a finite number of subtriangles such that any two subtriangles are either disjoint or else have a vertex or a side in common. Moreover, suppose the vertices of the subtriangles are all denoted by the letters A, B, and C in such a way that the letter A never appears on the side BC of the original triangle, the letter B never appears on the side AC and the letter C never appears on the side AB. Then there is at least one subtriangle whose vertices are denoted by different letters.*

Proof.   First we note that if an interval AB is divided into a finite number of subintervals, where each point of division is denoted by either A or B, then the number of subintervals with end points denoted by different letters is odd. In fact, let p_{AA} be the number of subintervals with both end points denoted by A, let p_{AB} be the number of subintervals with one end point denoted by A and the other by B, and let p'_A be the number of points of division denoted by A. Examining each subinterval in turn, independently of the others, we find that the end point A appears a number of times equal to $2p_{AA} + p_{AB}$ (why?). But this number must also equal $2p'_A + 1$, since each "interior" end point A is counted twice. Therefore

$$2p_{AA} + p_{AB} = 2p'_A + 1,$$

and hence p'_A is odd.

---

[25] Such a mapping is said to be *one-to-one and continuous in both directions*.

Turning to the case of the triangle ABC, let p_{AAB} be the number of subtriangles with vertices A, A and B (in any order), and similarly for p_{ABB} and p_{ABC}. Moreover, let p'_{AB} be the number of sides of subtriangles with end points A and B which lie inside the original triangle, and let p''_{AB} be the number of sides of subtriangles with end points A and B which lie on the boundary of the original triangle. Examining the sides of each subtriangle in turn, independently of the others, we find that the side AB appears a number of times equal to $2p_{AAB} + 2p_{ABB} + p_{ABC}$ (why?). This number must also equal $2p'_{AB} + p''_{AB}$, since each "interior" side AB is counted twice, and therefore

$$2p_{AAB} + 2p_{ABB} + p_{ABC} = 2p'_{AB} + p''_{AB}. \qquad (7.31)$$

But "boundary" sides AB can only lie on the side AB of the original triangle, and hence p''_{AB} is odd, by the argument given at the beginning of the proof. It follows from (7.31) that p_{ABC} is also odd. Thus $p_{ABC} \geqslant 1$, and the proof is complete.

We are now in a position to prove a theorem widely used in the theory of differential equations and other branches of mathematics:

THEOREM (*Brouwer's fixed point theorem*).[26]   *Let $f(x)$ be a continuous mapping of an n-dimensional ball K into itself ($n \geqslant 1$). Then $f(x)$ carries at least one point $a \in K$ into itself, i.e., $f(a) = a$.*

Proof. If $n = 1$, the theorem is an immediate consequence of the intermediate value theorem for continuous functions (why?). We now give the proof for $n = 2$, where K is a disk, leaving the case of arbitrary n to the problems.

First we note that if Brouwer's theorem holds for any set K' homeomorphic to K, then it also holds for K. In fact, let f be a continuous mapping of K into itself and let φ be a homeomorphic mapping of K' onto K. Then $\varphi^{-1} f \varphi$ is a continuous mapping of K' into itself, and hence has a fixed point a by hypothesis:

$$\varphi^{-1}\{f[\varphi(a)]\} = a.$$

But then

$$f[\varphi(a)] = \varphi(a),$$

i.e., $\varphi(a)$ is a fixed point of the mapping f.

This observation allows us to replace the disk K by a triangle ABC, and f by a continuous mapping of ABC into itself. Given any $\varepsilon_1 > 0$, we "triangulate" ABC, i.e., we divide ABC into subtriangles as in

---

[26] Proved by the Dutch mathematician L. E. J. Brouwer in 1910. An equivalent assertion was proved and applied to the theory of differential equations by the Latvian mathematician P. G. Bohl in 1904.

Sperner's lemma, all with sides of length less than ε_1. Suppose f carries no point of ABC into itself. Then we label the vertices of the subtriangles according to the following rule: A vertex is denoted by A if f carries it nearer to the side BC of the original triangle, by B if f carries it nearer to the side AC, and by C if f carries it nearer to the side BC. If f carries a vertex nearer to two sides of the original triangle, say BC and AC, the vertex can be denoted by either A or B. Then it is easy to see that this rule satisfies the conditions of Sperner's lemma, and hence there is at least one subtriangle, say $(ABC)_1$, whose vertices are denoted by different letters.

Now let $\varepsilon_1, \ldots, \varepsilon_m, \ldots$ be any sequence of positive numbers converging to zero, and carry out the above construction for every m. By going over to subsequences, we can assume without loss of generality that the corresponding triangles $(ABC)_m$ converge to a point a (why?). Since a is not fixed under the mapping f, it must move away from one of the sides, say BC, of the original triangle. But then every point of the triangle $(ABC)_m$ moves away from BC if n is sufficiently large. This contradicts the construction of $(ABC)_m$, thereby proving the theorem.

Problem 1. Generalize Sperner's lemma to n dimensions (proving that there is an odd number of subtetrahedra with appropriately labelled vertices), and then use mathematical induction to prove Brouwer's theorem for arbitrary n.

Problem 2. Prove Brouwer's theorem for arbitrary n by the following alternative method: First use the thoery of Sec. 57 to prove the theorem for sufficiently smooth mappings, and then approximate an arbitrary continuous mapping by smooth mappings.

Problem 3. Use Brouwer's theorem to show that no closed bounded n-dimensional set with interior points can be continuously mapped onto its boundary in such a way that every boundary point is carried into itself.

59. Applications of Brouwer's Theorem

First we generalize the corollary to Theorem 1, p. 186 on critical points inside a closed trajectory:

THEOREM 1. *Given an n-dimensional autonomous system* (7.2) *and a set K homeomorphic to an n-dimensional ball, suppose every trajectory of* (7.2) *beginning in K at* $t = 0$ *stays inside K for all* $t > 0$. *Then K contains at least one critical point of the system.*

Proof. Given any $\tau_1 > 0$, consider the mapping carrying an arbitrary point $x^0 \in K$ into the point $x(\tau_1; x^0)$. By hypothesis, there is a continuous mapping of K into itself, and hence, by Brouwer's theorem, there is a point $x^1 \in K$ such that $x(\tau_1; x^1) = x^1$, i.e., a trajectory beginning at x^1 returns to x^1 after a time τ_1. Similarly, if $\tau_1, \ldots, \tau_m, \ldots$ is any sequence of positive numbers converging to zero, there is a point $x^m \in K$ such that $x(\tau_m; x^m) = x^m$. By going over to subsequences, we can assume without loss of generality that the sequence $\{x^m\}$ converges to a point $\bar{x} \in K$.

We now show that $\bar{x}$ is a critical point, thereby completing the proof. Clearly

$$x\left(\left[\frac{t}{\tau_m}\right]\tau_m; x^m\right) = x^m \tag{7.32}$$

for any fixed t. The left-hand side of (7.32) approaches $x(t; \bar{x})$ as $m \to \infty$ (why?), while the right-hand side approaches $\bar{x}$. Therefore

$$x(t; \bar{x}) = \bar{x}$$

for all t, i.e., $\bar{x}$ is a critical point, as asserted.

Next we prove a result related to Theorem 2, p. 171 on cycles:[27]

THEOREM 2. *Given a three-dimensional autonomous system* (7.2), *let T be a torus such that every trajectory of* (7.2) *beginning inside T at* $t = 0$ *stays inside T for all* $t > 0$. *Moreover, suppose every point x moving along a trajectory in T, as described by the system* (7.2), *has positive angular velocity about the axis of* T.[28] *Then T contains at least one cycle.*

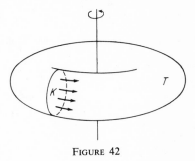

FIGURE 42

Proof. Let the disk K be a cross section of T, as in Figure 42, and let l_{x^0} be a trajectory beginning at a point $x^0 \in K$. By hypothesis, l_{x^0} stays inside T as t increases. The angular velocity of the points moving about the axis of T, being positive and continuous, has a positive minimum. Therefore, after a finite time, l_{x^0} intersects K again in a point x^1 depending on x^0:

$$x^1 = \varphi(x^0).$$

---

[27] The reader should try to generalize Theorem 2, which is a rather special result in three dimensions (chosen only to illustrate the technique of applying Brouwer's theorem).

[28] If the x_3-axis is chosen as the axis of the torus, the requirement on the angular velocity takes the form

$$x_1 \frac{dx_2}{dt} - x_2 \frac{dx_1}{dt} \equiv x_1 f_2(x_1, x_2, x_3) - x_2 f_1(x_1, x_2, x_3) > 0.$$

The function φ is a mapping of K into itself, playing the role of a succession function (see Sec. 55). Moreover, φ is continuous, because of the continuous dependence of the solution on the initial conditions. It follows from Brouwer's theorem that there is a point $\bar{x} \in K$ such that $\varphi(\bar{x}) = \bar{x}$. Thus there is a cycle passing through $\bar{x}$, and the proof is complete.

Problem 1. Let the set K, with boundary Γ, be homeomorphic to an n-dimensional ball, and suppose that for every point $a \in \Gamma$ there is an n-dimensional right circular cone P_a (including its base) with vertex at a such that

1) P_a is completely contained in K;

2) P_a depends continuously on a;

3) The cones P_a ($a \in \Gamma$) are all congruent.

Moreover, given an n-dimensional autonomous system (7.2), suppose the vector $f(a)$, applied at an arbitrary point $a \in \Gamma$, is never directed into P_a. Prove that K contains at least one critical point. In particular, use this fact to prove that if K is bounded by a smooth surface on which the velocity field is never directed along the exterior normal, then K contains at least one critical point.

Problem 2. Use Brouwer's theorem to find sufficient conditions for the existence of a periodic solution of the general system

$$\frac{dy_i}{dx} = f_i(x, y_1, \ldots, y_n), \qquad i = 1, \ldots, n,$$

if the functions f_i are all periodic in x with the same period.

Supplement

FIRST-ORDER PARTIAL
DIFFERENTIAL EQUATIONS

This supplement is devoted to the theory of first-order partial differential equations involving one unknown function. A basic feature of the theory is that all solutions of such equations reduce to integration of ordinary differential equations. Just how this reduction is accomplished will be described below.

60. Semilinear Equations

We begin by considering equations of the form

$$\sum_{i=1}^{n} a_i(x_1, \ldots, x_n) \frac{\partial u}{\partial x_i} + b(x_1, \ldots, x_n, u) = 0, \tag{1}$$

where it is assumed that

1) The coefficients $a_i(x_1, \ldots, x_n)$ have continuous first derivatives with respect to all their arguments on some domain G in $(x_1, \ldots, x_n)$ space;

2) The inequality

$$\sum_{i=1}^{n} a_i^2(x_1, \ldots, x_n) > 0$$

holds on G;

3) The function $b(x_1, \ldots, x_n, u)$ has continuous first derivatives with respect to all its arguments for $(x_1, \ldots, x_n) \in G$ and $|u| < M$.

In particular, condition 3 holds if $b(x_1, \ldots, x_n, u)$ is a linear function of u with coefficients which have continuous first derivatives with respect to all the x_i, and in this case equation (1) is called *linear*. In general, the unknown function u appears in $b(x_1, \ldots, x_n, u)$ in a nonlinear way, and then equation (1) is said to be *semilinear*.

Now consider the system of ordinary differential equations

$$\frac{dx_i}{ds} = \frac{a_i(x_1, \ldots, x_n)}{\sqrt{\sum\limits_{j=1}^{n} a_j^2(x_1, \ldots, x_n)}}, \qquad i = 1, \ldots, n. \tag{2}$$

Because of conditions 1 and 2, the right-hand sides of (2) have continuous first derivatives with respect to all the x_i. Hence there is one and only one integral curve of the system (2) passing through each point of G (the parameter s equals the arc length along the integral curve). These integral curves are called *characteristic curves* of equation (1).

THEOREM 1 (*Uniqueness theorem*). *If the function $u(x_1, \ldots, x_n)$ has continuous first derivatives and satisfies equation (1) on a domain G, then all the values of $u(x_1, \ldots, x_n)$ along an arc of a characteristic curve H, where $|u| < M$, are uniquely determined by its value at any point $(x_1^0, \ldots, x_n^0)$ of the arc.*

Proof. Dividing both sides of (1) by $\sqrt{a_1^2 + \cdots + a_n^2}$ and taking account of (2), we obtain

$$\sum_{i=1}^{n} \frac{a_i}{\sqrt{a_1^2 + \cdots + a_n^2}} \frac{\partial u}{\partial x_i} + \frac{b}{\sqrt{a_1^2 + \cdots + a_n^2}}$$

$$= \sum_{i=1}^{n} \frac{\partial u}{\partial x_i} \frac{dx_i}{ds} + \frac{b}{\sqrt{a_1^2 + \cdots + a_n^2}} = \frac{du}{ds} + \frac{b}{\sqrt{a_1^2 + \cdots + a_n^2}} = 0, \tag{3}$$

where the replacement of

$$\sum_{i=1}^{n} \frac{\partial u}{\partial x_i} \frac{dx_i}{ds}$$

by du/ds is justified by the assumed continuity of the first derivatives of u.[1] Let H be a characteristic curve passing through a point $(x_1^0, \ldots, x_n^0)$ of G, and suppose $|u(x_1^0, \ldots, x_n^0)| < M$. Then

$$x_i = \varphi_i(s, x_1^0, \ldots, x_n^0), \qquad i = 1, \ldots, n \tag{4}$$

along H, where φ_i and its first derivatives are continuous in s and all the x_j^0. Substituting (4) into

$$\frac{b}{\sqrt{a_1^2 + \cdots + a_n^2}},$$

---

[1] See e.g., D. V. Widder, *Advanced Calculus*, second edition, Prentice-Hall, Inc., Englewood Cliffs, N.J. (1961), p. 15.

we find that u satisfies the differential equation

$$\frac{du}{ds} = \psi(s, u, x_1^0, \ldots, x_n^0) \tag{5}$$

along H, where ψ has continuous first derivatives with respect to all its arguments if $|u| < M$ (cf. Sec. 21). Therefore the value of u at any point of an arc of H where $|u| < M$ is uniquely determined by its value at any point of the arc, in particular at the point $(x_1^0, \ldots, x_n^0)$.

THEOREM 2 (*Existence of a solution of the Cauchy problem*). *Let S be an $(n-1)$-dimensional surface lying in the domain G,[2] and suppose S has a continuously turning tangent plane. Suppose further that S is not tangent to any of the characteristic curves of equation* (1). *Let f be any function defined on S such that*

1) *The absolute value of f is less than M;*
2) *Each point of S has a neighborhood in which f can be represented as a function of $n-1$ of the coordinates $x_1, \ldots, x_n$, with continuous first derivatives with respect to these coordinates.*

Finally suppose that the surface S has a neighborhood R_0 such that

1) *R_0 is contained in G;*
2) *A characteristic curve passing through any point of the surface S does not intersect S again when continued in both directions inside R_0, and only one arc of a characteristic curve passes through each point of R_0;*
3) *Given any point $(x_1^0, \ldots, x_n^0)$ of the surface S, the solution of equation* (3) *satisfying the initial condition $u(0) = f(x_1^0, \ldots, x_n^0)$ can be continued along the entire arc of the characteristic curve inside R_0, without exceeding M in absolute value.[3]*

Then there exists a function $u(x_1, \ldots, x_n)$ defined on R_0 such that

1) *u has continuous partial derivatives with respect to all the x_i;*
2) *u is an "integral surface," i.e., u satisfies equation* (1);
3) *u satisfies the "initial data," i.e., u coincides with f on S.*

Proof. By solving the *Cauchy problem* for equation (1), we mean finding a function $u(x_1, \ldots, x_n)$ with the above properties. For simplicity, we confine ourselves to the case of two independent variables ($n = 2$), leaving it to the reader to extend the proof to the case $n > 2$. To avoid the use of subscripts, we write equation (1) for $n = 2$ in the form

$$a(x, y) \frac{\partial z}{\partial x} + b(x, y) \frac{\partial z}{\partial y} + c(x, y, z) = 0. \tag{6}$$

---

[2] S can be either closed or open; in the latter case, S is not regarded as including its boundary.

[3] It is assumed that $s = 0$ on S.

Correspondingly, equations (2)–(5) become

$$\frac{dx}{ds} = \frac{a(x, y)}{\sqrt{a^2 + b^2}}, \qquad \frac{dy}{ds} = \frac{b(x, y)}{\sqrt{a^2 + b^2}}, \qquad (7)$$

$$\frac{dz}{ds} + \frac{c(x, y, z)}{\sqrt{a^2 + b^2}} = 0, \qquad (8)$$

$$x = \varphi(s, x^0, y^0), \qquad y = \psi(s, x^0, y^0), \qquad (9)$$

$$\frac{dz}{ds} = \chi(s, z, x^0, y^0). \qquad (10)$$

The "surface" S on which f is defined is now a curve in the xy-plane. Thus, in this case, the Cauchy problem has the following geometric interpretation: Find an integral surface $z = z(x, y)$ satisfying equation

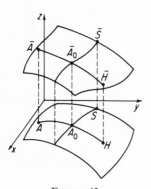

(6) and passing through a given spatial curve $\bar{S}$ whose projection onto the xy-plane is S (see Figure 43). Since the value of z along every characteristic curve H is uniquely determined by the value of z at the point in which H intersects S, we should be able to find $z(x, y)$ by examining all such characteristic curves.

We now pursue this idea in more detail. On every characteristic curve H intersecting S in a point A_0, we determine the function z which satisfies equation (10) and takes the same value at A_0 as the given function f. In general, the function z cannot be determined

FIGURE 43

in this way on the whole characteristic curve H, since in continuing z along H we may leave the domain of values of z where $c(x, y, z)$ is defined or we may intersect S twice. However, by hypothesis, we can determine z on a whole neighborhood R_0 of S. Thus all that remains is to prove that the function $z(x, y)$ so constructed has continuous derivatives with respect to x and y. Then the relation

$$\frac{\partial z}{\partial x} \frac{dx}{ds} + \frac{\partial z}{\partial y} \frac{dy}{ds} = \frac{dz}{ds}$$

will hold on R_0, and hence z will satisfy not only equation (10), but equations (8) and (6) as well.

Before proving the existence and uniqueness of the derivatives of z with respect to x and y at any point $A = (x, y)$ of the domain R_0, we introduce new curvilinear coordinates in a neighborhood of A as follows. Suppose the characteristic curve H drawn through A intersects S in the point $A_0 = (x^0, y^0)$. To be explicit, we assume that the tangent at A_0 to the curve S is not parallel to the y-axis (note that the argument given below remains in force if the roles of the x and y-axes are reversed). Then the arc of S near A_0 can be represented by an equation of the form

$$y^0 = F(x^0),$$

where the function F has a continuous derivative. On the other hand, since the functions $\varphi(s, x^0, y^0)$ and $\psi(s, x^0, y^0)$ have continuous derivatives with respect to s, x^0, and y^0 (cf. Sec. 29), the functions

$$\varphi[s, x^0, F(x^0)] \equiv \tilde{\varphi}(s, x^0), \qquad \psi[s, x^0, F(x^0)] \equiv \tilde{\psi}(s, x^0)$$

have continuous derivatives with respect to s and x^0. Let s and x^0 be the new coordinates. Then, as we now show, the Jacobian of the functions $x = \tilde{\varphi}(s, x^0)$, $y = \tilde{\psi}(s, x^0)$ is nonvanishing. In fact, these functions satisfy the system of ordinary differential equations (7), which for brevity we write as

$$\frac{dx}{ds} = \Phi(x, y), \qquad \frac{dy}{ds} = \Psi(x, y).$$

Substituting $\tilde{\varphi}$ and $\tilde{\psi}$ for x and y, we obtain the following identities in s and x^0:

$$\frac{d\tilde{\varphi}(s, x^0)}{ds} \equiv \Phi[\tilde{\varphi}(s, x^0), \tilde{\psi}(s, x^0)],$$

$$\frac{d\tilde{\psi}(s, x^0)}{ds} \equiv \Psi[\tilde{\varphi}(s, x^0), \tilde{\psi}(s, x^0)]. \tag{11}$$

Since the functions $\tilde{\varphi}$, $\tilde{\psi}$, Φ, and Ψ have continuous derivatives with respect to all their arguments, the right-hand sides of the identities (11) have continuous derivatives with respect to s and x^0, and hence the same is true of the left-hand sides. Therefore, differentiating both sides of (11) with respect to s, x^0, and setting

$$\frac{\partial \tilde{\varphi}}{\partial s} = D_0 \tilde{\varphi}, \quad \frac{\partial \tilde{\psi}}{\partial s} = D_0 \tilde{\psi}, \quad \frac{\partial \tilde{\varphi}}{\partial x^0} = D_1 \tilde{\varphi}, \quad \frac{\partial \tilde{\psi}}{\partial x^0} = D_1 \tilde{\psi},$$

we obtain[4]

$$\frac{d D_p \tilde{\varphi}}{ds} \equiv \frac{\partial \Phi}{\partial \tilde{\varphi}} D_p \tilde{\varphi} + \frac{\partial \Phi}{\partial \tilde{\psi}} D_p \tilde{\psi},$$

$$\frac{d D_p \tilde{\psi}}{ds} \equiv \frac{\partial \Psi}{\partial \tilde{\varphi}} D_p \tilde{\varphi} + \frac{\partial \Psi}{\partial \tilde{\psi}} D_p \tilde{\psi}, \qquad (p = 0, 1).$$

---

[4] Here we rely on the following theorem (see e.g., T. M. Apostol, *op. cit.*, p. 121): Given a function f defined on a domain G in the xy-plane, suppose the partial derivatives f_x, f_y and f_{xy} exist and are continuous on G. Then f_{yx} exists everywhere in G, and is equal to f_{xy}.

Since the coefficients $\partial\Phi/\partial\tilde{\varphi}, \ldots, \partial\Psi/\partial\tilde{\psi}$ do not depend on p, the functions $D_p\tilde{\psi}$ and $D_p\tilde{\psi}$ satisfy the same system of linear differential equations for both $p = 0$ and $p = 1$. Therefore the Jacobian

$$W = \begin{vmatrix} \dfrac{\partial\tilde{\varphi}}{\partial x^0} & \dfrac{\partial\tilde{\varphi}}{\partial s} \\[2ex] \dfrac{\partial\tilde{\psi}}{\partial x^0} & \dfrac{\partial\tilde{\psi}}{\partial s} \end{vmatrix},$$

which is a Wronskian, is nonzero on the whole arc H if and only if it is nonzero at point A_0 where the characteristic curve intersects S (cf. Sec. 33). But at this point

$$W = \begin{vmatrix} 1 & \dfrac{\partial\tilde{\varphi}}{\partial s} \\[2ex] \dfrac{dF}{dx^0} & \dfrac{\partial\tilde{\psi}}{\partial s} \end{vmatrix} = \dfrac{\partial\tilde{\psi}}{\partial s} - \dfrac{dF}{dx^0}\dfrac{\partial\tilde{\varphi}}{\partial s},$$

where the expression on the right is the cosine of the angle between the normal at A_0 to the curve S and the tangent to the characteristic curve H at the same point A_0, multiplied by a nonzero factor (why?). By hypothesis, this cosine is nonzero, and hence the Jacobian W is nonzero everywhere on H.

Therefore, by the implicit function theorem, the system of equations

$$x = \tilde{\varphi}(s, x^0), \qquad y = \tilde{\psi}(s, x^0)$$

can be solved for s and x^0 in a neighborhood of H. Moreover, since $\tilde{\varphi}$ and $\tilde{\psi}$ have continuous derivatives with respect to s and x^0, the implicitly defined functions s and x^0 will have continuous derivatives with respect to x and y.[5] Hence to prove that the function z constructed earlier on the domain R_0 has continuous derivatives with respect to x and y on R_0, we need only show that if z is regarded as a function of s and x^0 in a neighborhood of H, then it has continuous derivatives with respect to s and x^0. But this can be seen as follows: Substituting $F(x^0)$ for y^0 in the right-hand side of equation (10), we find that the function z satisfies an equation of the form

$$\dfrac{dz}{ds} = \tilde{\chi}(s, z, x^0),$$

where the function $\tilde{\chi}$ has continuous derivatives with respect to all its arguments. Moreover, by hypothesis, the function f giving the initial

---

[5] See e.g., T. M. Apostol, *op. cit.*, p. 147.

values of the function z for $s = 0$ (on the curve S) has a continuous derivative with respect to x^0. Therefore, using the theorem of Sec. 21, we find that z has continuous derivatives with respect to s and x^0 in a neighborhood of H, thereby completing the proof of Theorem 2.

Remark 1. Suppose the surface S and the function f defined on S satisfy the first two conditions of Theorem 2. Then it is easy to verify the existence of a neighborhood R_0 satisfying the conditions of the theorem. In fact, we need only choose the set R_0 formed from sufficiently small arcs of characteristic curves passing through points of S. This guarantees the existence of a solution of the Cauchy problem in a sufficiently "thin" neighborhood of S.

Remark 2. Unless it is assumed that the functions a_k and b have continuous derivatives, equation (1) may not have a solution with continuous derivatives. For example, following N. M. Gyunter, consider the equation

$$\frac{\partial z}{\partial x} + \frac{\partial z}{\partial y} = b(x - y), \tag{12}$$

where $b(w)$ is Weierstrass' continuous nowhere differentiable function.[6] We introduce new variables u, v and w by setting

$$x + y = v, \quad x - y = w, \quad z(x, y) = u(v, w).$$

Suppose that on some domain G of the xy-plane there exists a solution $z(x, y)$ of equation (12) which has continuous derivatives with respect to x and y.[7] In terms of the new variables, (12) becomes

$$\frac{\partial u}{\partial v} = \frac{1}{2} b(w).$$

The solutions of this equation are all given by the formula

$$u(v, w) = \frac{1}{2} b(w)v + c(w),$$

where $c(w)$ is an arbitrary function of w, i.e.,

$$z(x, y) = \frac{1}{2} (x + y)b(x - y) + c(x - y). \tag{13}$$

But it is easy to see that there is no domain in the xy-plane where the function z given by (13) can have derivatives with respect to x and y. In fact, if z had

---

[6] See e.g., E. C. Titchmarsh, *The Theory of Functions*, second edition, Oxford University Press, London (1939), p. 351.

[7] By a *solution* of (12) is meant a function $z(x, y)$ with partial derivatives $\partial z/\partial x$ and $\partial z/\partial y$ everywhere in G which satisfy (12).

such derivatives at the points (x, y) and $(x + \varepsilon, y + \varepsilon)$, then so would the function

$$z(x + \varepsilon, y + \varepsilon) - z(x, y) = \varepsilon b(x - y),$$

which is impossible. In other words, $z(x, y)$ cannot satisfy equation (12), contrary to our original assumption.

It can be shown[8] that every *continuous* solution of equation (12) is of the form (13), even if it is not required that the solution have continuous derivatives. Thus we arrive at the conclusion that there is no domain in which equation (12) has a continuous solution.

Problem 1. Prove that if the initial data are specified on a characteristic curve, then equation (6) either has no solution or all or else has infinitely many solutions. When does each case occur?

Problem 2. Show that if $n = 2$ and if the domain is simply connected, then the solution of a linear equation can be continued onto any domain consisting of points of characteristic curves drawn through S. Give an example showing that this is not always possible if the domain is not simply connected.

Comment. If $n > 2$, the continuation is not always possible even for a simply connected domain.

Problem 3. The theory becomes much simpler for generalized solutions of equation (1), by which we mean here continuous functions satisfying equation (3) along every characteristic curve. Consider generalized solutions of the equation

$$\frac{\partial z}{\partial x} = 0$$

in some domain R of the xy-plane, if the intitial data are specified on a piecewise smooth curve S contained in R. Find conditions for the existence of a (generalized) solution of the Cauchy problem, and conditions for the existence of a unique solution. Find conditions for the null solution to be the only solution corresponding to zero initial data. (How does this requirement differ from the preceding one?) Can we always infer the uniqueness of the solution in a smaller domain R from uniqueness of the solution in a larger domain containing R? Try to extend these results to the equation

$$a(x, y) \frac{\partial z}{\partial x} + b(x, y) \frac{\partial z}{\partial y} = 0 \qquad (a^2 + b^2 > 0),$$

where a and b are continuously differentiable functions. Give an example of such an equation in a domain R, whose solutions are all constants.

Problem 4. Give examples of cases where several characteristic curves go through certain points. Consider the second equation of Prob. 3 in the half-plane $0 \leqslant x < \infty$ with continuous Cauchy data on the x-axis, where a and b are continuous functions, $a \neq 0$ and it is assumed for simplicity that $\sup |b/a| < \infty$.

---

[8] R. Baire, *Sur les fonctions de variables réelles*, Annali di Mat. (3), **3**, 1 (1899).

Show that the (generalized) solution, if it exists, must be unique. Show that a solution exists if the characteristic curves can be uniquely continued in the direction of decreasing x, but otherwise a solution may not exist. What is a necessary and sufficient condition for existence of a solution of the Cauchy problem? What new features appear if b/a is not required to be bounded?

61. First Integrals of a System of Ordinary Differential Equations

According to the preceding section, the system consisting of the ordinary differential equations (2) and the equation

$$\frac{du}{ds} + \frac{b}{\sqrt{a_1^2 + \cdots + a_n^2}} = 0$$

(where s is a parameter) defines a family of integral curves in $(x_1, \ldots, x_n, u)$ space, which make up integral surfaces

$$u = u(x_1, \ldots, x_n)$$

of equation (1). These ordinary differential equations can be written in the symmetric form

$$\frac{dx_1}{a_1} = \frac{dx_2}{a_2} = \cdots = \frac{dx_n}{a_n} = \frac{du}{-b} \cdot \qquad (14)$$

It is sometimes easy to find functions

$$\varphi(x_1, \ldots, x_n, u)$$

which are not identically constant and have constant values along every integral curve of the system (14). Such functions are called *first integrals* of the system (14).

For example, consider equation (6) involving two independent variables. In this case, the family of integral curves making up the integral surfaces is determined by the system

$$\frac{dx}{a} = \frac{dy}{b} = \frac{dz}{-c} \cdot \qquad (15)$$

Suppose we have managed to find two first integrals

$$\varphi(x, y, z), \qquad \psi(x, y, z)$$

of the system (15) such that at least one minor of order two of the matrix

$$\left\| \begin{array}{ccc} \dfrac{\partial \varphi}{\partial x} & \dfrac{\partial \varphi}{\partial y} & \dfrac{\partial \varphi}{\partial z} \\[2mm] \dfrac{\partial \psi}{\partial x} & \dfrac{\partial \psi}{\partial y} & \dfrac{\partial \psi}{\partial z} \end{array} \right\|$$

is nonzero at every point of a domain G_z in (x, y, z) space.[9] Then the system of equations

$$\varphi(x, y, z) = \varphi(x^0, y^0, z^0), \qquad \psi(x, y, z) = \psi(x^0, y^0, z^0) \qquad (16)$$

determines a curve L in the domain G_z, since (16) determines two coordinates as functions of the third in a neighborhood of every point where both equations are satisfied. This curve will in general consist of several disjoint pieces. However, we shall assume that *in the domain G_z under consideration every curve determined by the system* (16) *consists of only one piece*. By the definition of a first integral, the function $\varphi(x, y, z)$ has the constant value $\varphi(x^0, y^0, z^0)$ along the integral curve L of the system (15) passing through the point (x^0, y^0, z^0), and similarly $\psi(x, y, z)$ has the constant value $\psi(x^0, y^0, z^0)$ along L. Therefore the curve determined by (16) must coincide with L in G_z. It follows that the system of equations

$$\varphi(x, y, z) = C_1, \qquad \psi(x, y, z) = C_2 \qquad (17)$$

represents a complete integral of the system (15) in the domain G_z, since by choosing the values $\varphi(x^0, y^0, z^0)$ and $\psi(x^0, y^0, z^0)$ for the constants C_1 and C_2, we obtain the integral curve of the system (15) passing through any point (x^0, y^0, z^0) of G_z, i.e., we obtain any integral curve of the system.

We now consider the problem of finding the integral surface of equation (6) passing through the curve

$$x = \alpha(v), \quad y = \beta(v), \quad z = \gamma(v), \qquad (18)$$

where it is assumed that the functions α, β, γ are continuously differentiable and that the determinant

$$\begin{vmatrix} \dfrac{d\alpha}{dv} & a(\alpha, \beta) \\[2mm] \dfrac{d\beta}{dv} & b(\alpha, \beta) \end{vmatrix}$$

never vanishes. Geometrically, the second condition means that the curve

$$x = \alpha(v), \qquad y = \beta(v)$$

in the xy-plane is never tangent to a characteristic curve of equation (6). Under these conditions, it follows from the basic theorems of Sec. 60 that in some neighborhood of the curve (18) there is one and only one integral surface of equation (6) passing through the curve (18). Since, as we have seen, this integral surface is made up of integral curves of the system (15), passing

---

[9] The analysis of the general system (14) requires n first integrals, and will be left to the reader.

through points of the curve (18), the parametric equations for the required integral surface (with parameter v) can be found by substituting $\alpha(v)$, $\beta(v)$ and $\gamma(v)$ for x^0, y^0 and z^0 in the right-hand sides of the equations (16):

$$\varphi(x, y, z) = \varphi[\alpha(v), \beta(v), \gamma(v)] = \Phi(v),$$
$$\psi(x, y, z) = \psi[\alpha(v), \beta(v), \gamma(v)] = \Psi(v).$$

Remark. Every first integral $\chi(x, y, z)$ of the system (15) in the domain G_z is a function of $\varphi(x, y, z)$ and $\psi(x, y, z)$. In fact, by the definition of a first integral, $\chi(x, y, z)$ must have a constant value along every integral curve of (15). But, as shown above, such a curve is uniquely determined by the values which the functions $\varphi(x, y, z)$ and $\psi(x, y, z)$ take on the curve.

Example. *Find the integral surface of the equation*

$$2 \frac{\partial z}{\partial x} + 3 \frac{\partial z}{\partial y} + 5 = 0 \tag{6'}$$

passing through the curve

$$x = a_1 v, \quad y = a_2 v, \quad z = a_3 v,$$

where the constants a_1, a_2 and a_3 are such that

$$\begin{vmatrix} a_1 & 2 \\ a_2 & 3 \end{vmatrix} \neq 0. \tag{19}$$

The system (15) now takes the form

$$\frac{dx}{2} = \frac{dy}{3} = \frac{dz}{-5}. \tag{15'}$$

Integrating the equations

$$\frac{dx}{2} = \frac{dy}{3}$$

and

$$\frac{dx}{2} = \frac{dz}{-5},$$

we obtain the following first integrals of the system (15'):

$$3x - 2y, \qquad 5x + 2z.$$

Since

$$\begin{vmatrix} \dfrac{\partial \varphi}{\partial x} & \dfrac{\partial \varphi}{\partial y} \\[2mm] \dfrac{\partial \psi}{\partial x} & \dfrac{\partial \psi}{\partial y} \end{vmatrix} = \begin{vmatrix} 3 & -2 \\ 5 & 0 \end{vmatrix} \neq 0,$$

the equations
$$3x - 2y = C_1, \qquad 5x + 2z = C_2$$

represent a complete integral of (15') in all of (x, y, z) space. For each pair of values of the constants C_1 and C_2, the system (7') determines a curve (a straight line) consisting of only one piece. Therefore the required integral of (6') is determined by the equations
$$3x - 2y = 3a_1v - 2a_2v,$$
$$5x + 2z = 5a_1v + 2a_3v.$$

To eliminate v from these equations, we solve the first equation for v [this is possible because of the condition (19')], and then substitute the resulting value of v into the second equation.

Problem 1. Show that in general, knowledge of k functionally independent first integrals of the system (4.4) allows us to reduce the number of unknown functions in the system by k.

Problem 2. In physical problems, first integrals often express conservation laws. Use the law of conservation of total energy, to find a first integral of the system
$$\frac{dx}{dt} = v, \qquad m\frac{dv}{dt} + bx = 0,$$

equivalent to the equation
$$m\frac{d^2x}{dt^2} + bx = 0$$

for free oscillations of a particle in the absence of friction.

62. Quasi-Linear Equations

We now consider equations of the form
$$\sum_{i=1}^{n} a_i(x_1, \ldots, x_n, u)\frac{\partial u}{\partial x_i} + b(x_1, \ldots, x_n, u) = 0, \qquad (20)$$

where it is assumed that

1) The functions $a_i(x_1, \ldots, x_n, u)$ and $b(x_1, \ldots, x_n, u)$ have continuous first derivatives with respect to all their arguments on some domain G in $(x_1, \ldots, x_n, u)$ space;

2) The inequality
$$\sum_{i=1}^{n} a_i^2(x_1, \ldots, x_n, u) > 0$$

holds on G.

Starting from a known solution $u(x_1, \ldots, x_n)$ of (20) with continuous first derivatives, we form the auxiliary system of ordinary differential equations

$$\frac{dx_i}{ds} = \frac{a_i[x_1, \ldots, x_n, u(x_1, \ldots, x_n)]}{\sqrt{\displaystyle\sum_{j=1}^{n} a_j^2[x_1, \ldots, x_n, u(x_1, \ldots, x_n)]}}, \quad i = 1, \ldots, n, \quad (21)$$

where the parameter s equals the arc length along the projection of the integral curve onto the plane $u = 0$. Substituting $u(x_1, \ldots, x_n)$ into (20) and dividing both sides of the resulting identity by

$$\sqrt{\sum_{j=1}^{n} a_j^2[x_1, \ldots, x_n, u(x_1, \ldots, x_n)]},$$

we find that

$$\sum_{i=1}^{n} \frac{a_i(x_1, \ldots, x_n, u)}{\sqrt{\displaystyle\sum_{j=1}^{n} a_j^2(x_1, \ldots, x_n, u)}} \frac{\partial u}{\partial x_i} + \frac{b(x_1, \ldots, x_n, u)}{\sqrt{\displaystyle\sum_{j=1}^{n} a_j^2(x_1, \ldots, x_n, u)}}$$

$$= \sum \frac{\partial u}{\partial x_i} \frac{dx_i}{ds} + \frac{b(x_1, \ldots, x_n, u)}{\sqrt{\displaystyle\sum_{j=1}^{n} a_j^2(x_1, \ldots, x_n, u)}} \quad (22)$$

$$= \frac{du}{ds} + \frac{b(x_1, \ldots, x_n, u)}{\sqrt{\displaystyle\sum_{j=1}^{n} a_j^2(x_1, \ldots, x_n, u)}} = 0.$$

Thus, in the case where the a_i depend on u, the expression

$$\sum_{i=1}^{n} \frac{a_i}{\sqrt{\displaystyle\sum_{j=1}^{n} a_j^2}} \frac{\partial u}{\partial x_i}$$

can still be represented as the derivative of u along some direction, but the direction now depends on u as well as on $x_1, \ldots, x_n$. Together (21) and (22) constitute a system of ordinary differential equations

$$\frac{dx_i}{ds} = \frac{a_i(x_1, \ldots, x_n, u)}{\sqrt{\displaystyle\sum_{j=1}^{n} a_j^2(x_1, \ldots, x_n, u)}}, \quad i = 1, \ldots, n,$$

$$\frac{du}{ds} = -\frac{b(x_1, \ldots, x_n, u)}{\sqrt{\displaystyle\sum_{j=1}^{n} a_j^2(x_1, \ldots, x_n, u)}}, \quad (23)$$

whose integral curves

$$x_i = x_i(s, x_1^0, \ldots, x_n^0, u^0), \qquad i = 1, \ldots, n,$$

$$u = u(s, x_1^0, \ldots, x_n^0, u^0)$$

are called *characteristic curves* of equation (20).[10]

The statement of the Cauchy problem for equation (20) is the same as for equation (1): *Find a solution of equation* (20) *taking given values on some* $(n - 1)$-*dimensional surface S in* $(x_1, \ldots, x_n)$ *space*, or more generally, *find an n-dimensional integral surface T of equation* (20) *passing through a given* $(n - 1)$-*dimensional surface $\bar{S}$ in* $(x_1, \ldots, x_n, u)$ *space*.[11] First we prove the uniqueness of such a solution, assuming that the solution is continuously differentiable and that the surface $\bar{S}$ has no tangent lines whose projections onto the plane $u = 0$ have the direction cosines specified by (21).[12] Since the initial values (for $s = 0$) of the x_i and u are known at every point $(x_1^0, \ldots, x_n^0, u^0)$ of the surface $\bar{S}$, the system (23) uniquely determines the x_i and u as functions of s, i.e., given any sufficiently small piece of $\bar{S}$, u is uniquely determined in the neighborhood of the projection of this piece onto the plane $u = 0$ which is covered by the projections of the characteristic curves intersecting $\bar{S}$. The "width" of this neighborhood never "shrinks to zero" since

1) $\bar{S}$ *has no tangents parallel to the u-axis*;[13]

2) By hypothesis, $\bar{S}$ *has no tangent lines whose projections onto the plane* $u = 0$ *have the direction cosines specified by* (21).

Thus any two integral surfaces passing through $\bar{S}$ coincide in some neighborhood of $\bar{S}$, and the uniqueness is proved.

To prove the existence of a solution of the Cauchy problem for equation (20), we consider the case of two independent variables x and y, as in Secs. 60 and 61, denoting the unknown function by $z = z(x, y)$. Then equations (20)

---

[10] Since semilinear equations are a special case of quasi-linear equations, there are two concepts of a characteristic curve for semilinear equations, one in the sense of Sec. 60, the other in the present sense. It is easily verified that the characteristic curves in the first sense are the projections of the characteristic curves in the second sense onto the plane $u = 0$.

[11] The surface T must represent a single-valued function $u(x_1, \ldots, x_n)$ satisfying equation (20) sufficiently near any of its points, but T may still be a multiple-valued function of $x_1, \ldots, x_n$ "in the large." For example, suppose T is that part of a helicoid with axis along the u-axis which lies outside some neighborhood of the axis.

[12] The direction cosines of the tangent lines to $\bar{S}$ are found by substituting the point of tangency into the right-hand sides of (21).

[13] This follows from the fact that the derivatives $\partial u / \partial x_i$ are assumed to exist on T, and existence implies finiteness (cf. p. 3).

and (23) become

$$a(x, y, z)\frac{\partial z}{\partial x} + b(x, y, z)\frac{\partial z}{\partial y} + c(x, y, z) = 0 \qquad (a^2 + b^2 > 0), \quad (24)$$

$$\frac{dx}{ds} = \frac{a(x, y, z)}{\sqrt{a^2 + b^2}}, \qquad \frac{dy}{ds} = \frac{b(x, y, z)}{\sqrt{a^2 + b^2}}, \qquad \frac{dz}{ds} = -\frac{c(x, y, z)}{\sqrt{a^2 + b^2}}, \quad (25)$$

while the characteristic curves and the curve $\bar{S}$ are given by

$$x = \varphi(s, x^0, y^0, z^0), \quad y = \psi(s, x^0, y^0, z^0), \quad z = \chi(s, x^0, y^0, z^0) \quad (26)$$

and

$$x^0 = \alpha(v), \quad y^0 = \beta(v), \quad z^0 = \gamma(v), \qquad (\alpha'^2 + \beta'^2 + \gamma'^2 > 0), \quad (27)$$

where v is a parameter. We shall assume that all given functions are continuously differentiable and that the curve $\bar{S}$ lies in the domain of definition of the coefficients a, b, c. Moreover, if we are interested in finding a single-valued solution rather than constructing an integral surface, it must also be assumed that the projection S of the curve $\bar{S}$ onto the plane $u = 0$ has no multiple points. In this case, the two italicized conditions take the form

$$\alpha'^2 + \beta'^2 > 0, \qquad \alpha'b - \beta'a \neq 0 \text{ along } \bar{S}. \quad (28)$$

We now verify that equations (26) and (27) determine a surface T in some three-dimensional neighborhood of the curve $\bar{S}$ such that

1) T has an equation of the form $z = z(x, y)$ near each of its points;

2) The function $z(x, y)$ is continuously differentiable.

Then it follows at once from (25), read in the reverse order, that $z(x, y)$ satisfies (24), and it is obvious that the initial conditions are satisfied. By the implicit function theorem, it is enough to show that

$$\frac{\partial(x, y)}{\partial(s, v)} \neq 0$$

on $\bar{S}$, since then we can solve the first two equations (26) for s and v in a neighborhood of $\bar{S}$, and afterwards substitute the results into the third equation.[14] But on $\bar{S}$, i.e., for $s = 0$, we have $x = \alpha(v)$, $y = \beta(v)$, and hence

$$\frac{\partial(x, y)}{\partial(s, v)} = \begin{vmatrix} \dfrac{\partial\varphi}{\partial s} & \dfrac{\partial\psi}{\partial s} \\ \dfrac{\partial\varphi}{\partial v} & \dfrac{\partial\psi}{\partial v} \end{vmatrix}_{s=0} = \begin{vmatrix} \dfrac{a}{\sqrt{a^2 + b^2}} & \dfrac{b}{\sqrt{a^2 + b^2}} \\ \alpha'(v) & \beta'(v) \end{vmatrix} = \frac{a\beta' - b\alpha'}{\sqrt{a^2 + b^2}} \neq 0,$$

---

[14] The functions φ, ψ and χ are continuously differentiable because of the continuous dependence of the solution of the system (25) on the initial data.

because of (25) and (28). This proves the existence of a solution of the Cauchy problem.

It should be emphasized that the existence of a solution is guaranteed only in a sufficiently small neighborhood of the curve S, which is not known in advance. The fact that this is an essential feature of the problem will be made clear in the next section.

63. Generalized Solutions of Linear and Quasi-Linear Equations

In Sec. 60 we constructed a solution of the Cauchy problem for a first-order linear partial differential equation, and then proved the uniqueness of the solution, assuming sufficient smoothness of the surface S, the function f defined on S and all the coefficients of the equation. Examples like the one at the end of Sec. 60 show that a solution of the problem may fail to exist if these assumptions are weakened. Thus, in many physical problems, it is important to extend the concept of a solution of the Cauchy problem in such a fashion that a solution will exist and be unique, under weaker conditions on the smoothness of f and the coefficients of the equation. We now describe one possible way in which this can be done.[15]

For simplicity, we consider the Cauchy problem for the case of two independent variables t and x, and a linear equation of the form

$$L(u) \equiv \frac{\partial u}{\partial t} + a(t, x) \frac{\partial u}{\partial x} + b(t, x)u + c(t, x) = 0, \qquad (29)$$

where a, b and c are continuously differentiable functions of t and x. The initial data will be specified on a segment of the line $t = 0$:

$$u(0, x) = f(x). \qquad (30)$$

Clearly, the line $t = 0$ is not tangent to any characteristic curves of equation (29). If a function $u(t, x)$ satisfies (29) on some domain G, then obviously

$$\iint_G L(u)F \, dt \, dx = 0, \qquad (31)$$

where F is any continuously differentiable function vanishing in a neighborhood of the boundary of G. Conversely, if $u(x, t)$ is continuously differentiable

---

[15] Concerning generalized solutions of linear and quasi-linear first-order partial differential equations, see e.g., R. Courant and D. Hilbert, *Methods of Mathematical Physics, Vol. 2*, Interscience Publishers, New York (1962), Chap. 2, Appendix 2 and Chap. 5, Sec. 9; O. A. Oleinik, *Discontinuous solutions of nonlinear differential equations* (in Russian), Uspekhi Mat. Nauk, **12**, no. 3, 3 (1957); I. G. Petrovski, *Lectures on Partial Differential Equations* (translated by A. Shenitzer), Interscience Publishers, New York (1954), Chap. 2, Sec. 9.

on G and if equation (31) holds for an arbitrary function of the above type, then $L(u) = 0$, i.e., $u(t, x)$ is a solution of equation (29) on G. In fact, if $L(u) \neq 0$ at a point (t^0, x^0), then $L(u)$ does not change sign in some neighborhood Ω of (t^0, x^0). But this contradicts (31) if we choose F to be positive in Ω and zero outside Ω.

We now write equation (31) in another form, using the identity

$$L(u)F = L^*(F)u + cF + \frac{\partial}{\partial t}(uF) + \frac{\partial}{\partial x}(auF), \qquad (32)$$

where, by definition,

$$L^*(F) \equiv -\frac{\partial F}{\partial t} - \frac{\partial}{\partial x}(aF) + bF.$$

Integrating (32) over G, using Green's theorem to transform the integral of the last two terms, and bearing in mind that F vanishes in a neighborhood of the boundary of G, we find that

$$\iint_G L(u)F \, dt \, dx = \iint_G [L^*(F)u + cF] \, dt \, dx.$$

The equation

$$\iint_G [L^*(F)u + cF] \, dt \, dx = 0 \qquad (33)$$

will serve as the starting point for our definition of a generalized solution of equation (29). It is important to note that (33) makes sense for nondifferentiable functions $u(t, x)$, and even for discontinuous $u(t, x)$. Here it will be assumed that discontinuities of $u(t, x)$ and its first derivatives $u_t = \partial u/\partial t$ and $u_x = \partial u/\partial x$ can occur only on a finite number of curves $x = \varphi(t)$, sharing no points except possibly end points, where $\varphi(t)$ is continuously differentiable. It will also be assumed that $u[t, \varphi(t) - \varepsilon]$ and $u[t, \varphi(t) + \varepsilon]$ approach uniform limits as $\varepsilon \to 0+$ on any arc of the indicated curves, and similarly for u_t and u_x. Such functions $u(t, x)$ will be called *piecewise continuous*.

> DEFINITION. *A piecewise continuous function $u(t, x)$ is said to be a generalized solution of equation* (29) *on the domain G if equation* (33) *holds for every continuously differentiable function $F(t, x)$ equal to zero in a neighborhood of the boundary of G.*

It is easy to see that a generalized solution $u(t, x)$ of equation (29) can have discontinuities only along characteristic curves. In fact, suppose $x = \varphi(t)$ is a curve of discontinuity of $u(t, x)$ and let $F(t, x)$ be any continuously differentiable function equal to zero outside a sufficiently small neighborhood Ω of the curve $x = \varphi(t)$. We can assume that $u(t, x)$ is continuously differentiable on each of the domains Ω_1 and Ω_2 into which the curve $x = \varphi(t)$ divides Ω, and hence $u(t, x)$ satisfies (29) on Ω_1 and Ω_2. Using (33), we find that

$$0 = \iint_\Omega [L^*(F)u + cF] \, dt \, dx$$

$$= \iint_{\Omega_1} [L^*(F)u + cF] \, dt \, dx + \iint_{\Omega_2} [L^*(F)u + cF] \, dt \, dx$$

$$= \iint_{\Omega_1} L(u)F \, dt \, dx + \iint_{\Omega_2} L(u)F \, dt \, dx + \int_{x=\varphi(t)} [u]F \, dx - [u]aF \, dt$$

$$= \int_{x=\varphi(t)} [u] \left\{ \frac{d\varphi}{dt} - a[t, \varphi(t)] \right\} F \, dt,$$

where the last two integrals are taken along the curve $x = \varphi(t)$, and

$$[u] = u[t, \varphi(t) - 0] - u[t, \varphi(t) + 0].$$

Since F is an arbitrary function on the curve $x = \varphi(t)$, vanishing of the last integral implies

$$\frac{d\varphi}{dt} - a = 0,$$

i.e., $x = \varphi(t)$ is a characteristic curve of equation (29). In the same way, the derivatives u_t and u_x can have discontinuities only along characteristic curves. In fact, suppose $u(t, x)$ is continuous along a curve $x = \varphi(t)$ which is not a characteristic curve, and let Ω_1 and Ω_2 be the same as before. Then by Theorems 1 and 2 of Sec. 60, $u(t, x)$ coincides both in $\bar{\Omega}_1$ and $\bar{\Omega}_2$ (the overbar denotes the closure) with the unique ordinary continuously differentiable solution, equal to $u[t, \varphi(t)]$ on the curve $x = \varphi(t)$. It follows that u_t and u_x can have no discontinuities on $x = \varphi(t)$. Moreover, it can easily be verified that any piecewise continuous function $u(t, x)$ is a generalized solution if it satisfies equation (29) at points of continuous differentiability and has discontinuities only on characteristic curves.

By a *generalized solution of the Cauchy problem* for equation (29) and the initial condition (30) on the domain G, we mean a generalized solution $u(t, x)$ of equation (29) which coincides with the given function $f(x)$ at continuity points. It is easy to see that such a function satisfies the equation

$$\iint_{G_1} [L^*(F)u + cF] \, dt \, dx - \int_{a_1}^{a_2} F(0, x)f(x) \, dx,$$

where G_1 is the intersection of G with the half-plane $t \geqslant 0$, and the segment $[a_1, a_2]$ is the intersection of the closure of G with the line $t = 0$. This equation can be taken as the definition of the generalized solution of the Cauchy problem for equation (29) and the initial condition (30) on the domain G.

An important property of linear partial differential equations, proved in Sec. 60, is the fact that the neighborhood R_0 of the surface S, in which a solution of the Cauchy problem is guaranteed to exist, does not depend on the function f specified on S. It can be shown that a generalized solution of the

Cauchy problem for equation (29) and the initial condition (30) exists in this same neighborhood of the line $t = 0$ and moreover is unique, provided that $f(x)$ is continuously differentiable everywhere except at a finite number of points where it and its derivatives have discontinuities of the first kind. This fact can be proved by the same method used in Sec. 60 to prove the existence and uniqueness of a solution of the Cauchy problem.

Next we consider quasi-linear equations of the first order. In this case, examples show that unlike linear equations, the domain of definition of solutions of the Cauchy problem depends on the function f and its derivatives. For example, for the equation

$$\frac{\partial u}{\partial t} + u \frac{\partial u}{\partial x} = 0, \tag{34}$$

the solution of the Cauchy problem satisfying the condition

$$u(0, x) = -\tanh \frac{x}{\varepsilon}, \qquad \varepsilon > 0$$

is defined only for $t < \varepsilon$, since it is not hard to show that the projections onto the tx-plane of the characteristic curves corresponding to the initial function $-\tanh (x/\varepsilon)$ intersect for $t > \varepsilon$ and "carry different values of u to the points of intersection."

By analogy with linear equations, we now introduce the concept of generalized solutions of first-order quasi-linear equations, confining our discussion to equations of the form[16]

$$\frac{\partial u}{\partial t} + \frac{\partial a(t, x, u)}{\partial x} + b(t, x, u) = 0. \tag{35}$$

By a *generalized solution* of equation (35) on a domain G we mean a piecewise smooth function $u(t, x)$ such that the equation

$$\iint_G \left[\frac{\partial F}{\partial t} u + \frac{\partial F}{\partial x} a(t, x, u) - b(t, x, u)F \right] dt \, dx = 0 \tag{36}$$

holds for every continuously differentiable function $F(t, x)$ equal to zero in a neighborhood of the boundary of G. The generalized solution of equation (35) satisfying the initial condition (30) at continuity points is called the *generalized solution of the Cauchy problem* (35), (30). Just as for equation (29), it can be shown that the generalized solution $u(t, x)$ of equation (35) satisfies the condition

$$\frac{d\varphi(t)}{dt} = \frac{a[t, x + 0, u(t, x + 0)] - a[t, x - 0, u(t, x - 0)]}{u(t, x + 0) - u(t, x - 0)} \tag{37}$$

---

[16] Because of their physical significance, equations of the form (35) are sometimes called *conservation laws*. The concept of generalized solutions of quasi-linear equations plays an important role in problems of gas dynamics (see the references cited in footnote 15).

on the curves of discontinuity $x = \varphi(t)$. It can also be shown that in general a curve of discontinuity of $u(t, x)$ is not the projection of a characteristic curve of equation (35).

Unlike the linear case, equation (36) does not uniquely determine a generalized solution of equation (35) satisfying the initial condition (30). For example, the generalized solution of the Cauchy problem for the equation

$$\frac{\partial u}{\partial t} + \frac{\partial(u^2/2)}{\partial x} = 0, \tag{34'}$$

satisfying the initial condition (30) where

$$f(x) = \begin{cases} 1 & \text{if } x < 0, \\ -1 & \text{if } x \geqslant 0, \end{cases} \tag{38}$$

is given by the following function $u_\alpha(t, x)$, defined for every $\alpha \geqslant 1$:

$$u_\alpha(t, x) = \begin{cases} 1 & \text{if } x \leqslant \dfrac{1-\alpha}{2} t, \\[2mm] -\alpha & \text{if } \dfrac{1-\alpha}{2} t < x \leqslant 0, \\[2mm] \alpha & \text{if } 0 < x \leqslant \dfrac{\alpha-1}{2} t, \\[2mm] -1 & \text{if } \dfrac{\alpha-1}{2} t < x. \end{cases} \tag{39}$$

The projections of the characteristic curves of $u_\alpha(t, x)$ onto the tx-plane are shown in Figure 44. Thus, to uniquely determine the generalized solution of the Cauchy problem for equation (35) on a domain lying in the half-plane $t > 0$, subject to the initial condition (30), (38), we need an extra condition, such as the requirement that the inequality

$$u(t, x - 0) > u(t, x + 0) \tag{40}$$

hold on the curves of discontinuity, in the case where $\partial^2 a/\partial u^2$ is nonnegative.[17] It is easy to see that (40) singles out one of the generalized solutions (39), i.e., the solution $u_1(t, x)$, which,

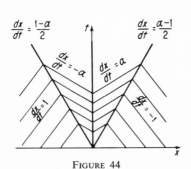

FIGURE 44

---

[17] For the proof, see O. A. Oleinik, *loc. cit.*

according to (39), is just

$$u_1(t, x) = \begin{cases} 1 & \text{if } x \leqslant 0, \\ -1 & \text{if } x > 0. \end{cases}$$

Remark 1. In the case of quasi-linear equations in several independent variables, conditions guaranteeing the uniqueness of a generalized solution of the Cauchy problem (for some large class of equations) have not yet been found.

Remark 2. Unlike the case of linear equations, a generalized solution of the Cauchy problem for the quasi-linear equation (35) cannot be constructed by using characteristic curves passing through the point $(0, x, u(0, x))$. For example, it is easy to see that the generalized solution of the Cauchy problem for equation (34) with the initial condition (38) is not uniquely determined by characteristic curves even in an arbitrarily small neighborhood of the line $t = 0$. In fact, suppose characteristic curves of equation (34), i.e., solutions of the system of equations

$$\frac{du}{dt} = 0, \qquad \frac{dx}{dt} = u,$$

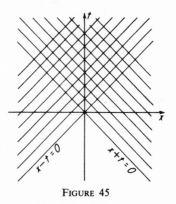

are drawn in (t, x, u) space through the points $(0, x, u(0, x))$, where $-\infty < x < +\infty$. In the case of a smooth function, the characteristic curves make up a surface corresponding to a solution of the Cauchy problem. However, if $u(0, x) = f(x)$, where $f(x)$ is given by (38), then all points of the tx-plane lying between the lines $x - t = 0$ and $x + t = 0$ are covered twice (with

FIGURE 45

different values of u) by the projections of the characteristic curves onto the tx-plane if $t > 0$, while the points between the same lines are not covered at all if $t < 0$ (see Figure 45).

Remark 3. There are other methods of constructing generalized solutions of quasi-linear equations. One of these, called the method of "small viscosity" because of its meaning in problems of gas dynamics, is of particular importance and is described in detail in Oleinik's paper (*loc. cit.*). In fact, the whole class of problems involving construction of generalized solutions of quasi-linear equations, continuation of the solutions for all $t > 0$ and conditions guaranteeing their uniqueness is of great importance in gas dynamics and other branches of mechanics. In gas dynamics, the curves of discontinuity of generalized solutions correspond to shock waves. Then conditions of the form (37) express the conservation of mass, energy and momentum in passing through a shock wave, while (40) expresses the fact that the entropy can only increase.

Problem 1. Define a generalized solution of the Cauchy problem for linear partial differential equations of the first order with any number of independent variables. Assuming that the initial function $f(x)$ is piecewise continuous, prove the appropriate existence and uniqueness theorem, and also derive the relations satisfied on surfaces of discontinuity of the generalized solutions.

Problem 2. Use equation (33) to define a continuous generalized solution of the Cauchy problem for equation (12) of Sec. 60, assuming that the initial function $f(x)$ is continuous. Prove the existence and uniqueness of this generalized solution.

Problem 3. Construct a generalized solution of the Cauchy problem for equation (34) in the domain $t > 0$ satisfying the condition (40), assuming that $f(x)$ is a piecewise constant function with finitely many discontinuities [first consider the case where $f(x)$ has one or two discontinuities]. Prove the uniqueness of the solution.

Problem 4. Prove that the condition (40) determines a unique generalized solution of the Cauchy problem for equation (34) in the half-plane $t \geqslant 0$.

64. Nonlinear Equations

We now consider nonlinear equations of the form

$$F\left(x_1, \ldots, x_n, u, \frac{\partial u}{\partial x_1}, \ldots, \frac{\partial u}{\partial x_n}\right) = 0, \tag{41}$$

where it is assumed that

1) The function F has continuous derivatives up to order 2 (inclusive) with respect to all its arguments in some $(2n + 1)$-dimensional domain G;

2) The inequality

$$\sum_{i=1}^{n} \left[\frac{\partial F}{\partial(\partial u/\partial x_i)}\right]^2 > 0 \tag{42}$$

holds on G.

For brevity, we introduce the notation

$$\frac{\partial F}{\partial x_i} = X_i, \quad \frac{\partial F}{\partial u} = U, \quad \frac{\partial u}{\partial x_i} = p_i, \quad \frac{\partial F}{\partial p_i} = P_i.$$

The Cauchy problem can be set up just as in Sec. 64. We begin by proving the uniqueness of the solution of the Cauchy problem, under certain conditions, at the same time developing tools which will be used to prove the existence of a solution.

Suppose $u(x_1, \ldots, x_n)$ is a solution of equation (41) with continuous second derivatives. Substituting this solution into (41), we differentiate the resulting identity with respect to each variable x_j, obtaining the equations

$$\sum_{i=1}^{n} P_i \frac{\partial p_i}{\partial x_j} + X_j + U p_j = \sum_{i=1}^{n} P_i \frac{\partial p_j}{\partial x_i} + X_j + U p_j = 0, \qquad (43)$$

which are quasi-linear in the p_i. Suppose we construct the integral curves in $(x_1, \ldots, x_n)$ space of the system

$$\frac{dx_i}{ds} = \frac{P_i}{\sqrt{\sum_{j=1}^{n} P_j^2}}, \qquad i = 1, \ldots, n, \qquad (44)$$

where the solution $u(x_1, \ldots, x_n)$ and its derivatives are substituted for u and p_j in the right-hand sides. Then equation (43) can be written in the form

$$\frac{dp_i}{ds} = -\frac{X_i + U p_i}{\sqrt{\sum_{j=1}^{n} P_j^2}}, \qquad i = 1, \ldots, n. \qquad (45)$$

Next we form the derivative of $u(x_1, \ldots, x_n)$ along the direction s defined by (44):

$$\frac{du}{ds} = \sum_{i=1}^{n} \frac{\partial u}{\partial x_i} \frac{dx_i}{ds} = \sum_{i=1}^{n} \frac{P_i p_i}{\sqrt{\sum_{j=1}^{n} P_j^2}}. \qquad (46)$$

The system consisting of equations (44), (45) and (46) uniquely defines x_i, p_i and u as functions of s, once their initial values are specified. The integral curves of this system in $(x_1, \ldots, x_n, u, p_1, \ldots, p_n)$ space are called *characteristic curves* of equation (41), and depend only on this equation.[18]

Now let S be the surface on which the initial values of u are specified, and suppose S has a continuously turning tangent plane. If S is given by the equations

$$x_i = x_i(v_1, \ldots, v_{n-1}), \qquad i = 1, \ldots, n,$$

then the function u can also be regarded as a function of $v_1, \ldots, v_{n-1}$ on S.

---

[18] The projections of these curves onto the plane $(x_1, \ldots, x_n)$ or onto the plane $(x_1, \ldots, x_n, u)$ are also sometimes called characteristic curves.

We shall assume that

1) *The functions $x_i(v_1, \ldots, v_{n-1})$ and $u(v_1, \ldots, v_{n-1})$ are continuous together with their partial derivatives up to order* 2;

2) *At least one minor of order $n - 1$ of the matrix*

$$\left\Vert \begin{array}{ccc} \dfrac{\partial x_1}{\partial v_1} & \cdots\cdots\cdots & \dfrac{\partial x_n}{\partial v_1} \\ \cdots\cdots\cdots\cdots\cdots\cdots \\ \dfrac{\partial x_1}{\partial v_{n-1}} & \cdots\cdots\cdots & \dfrac{\partial x_n}{\partial v_{n-1}} \end{array} \right\Vert$$

is nonzero at every point of S.

It is easy to see that the initial conditions determine the values of p_i on S, although admittedly not uniquely. In fact, the formula $u = u(x_1, \ldots, x_n)$ becomes an identity if we replace u and all the x_i by their expressions in terms of $v_1, \ldots, v_{n-1}$, and then differentiation of this identity gives

$$\sum_{i=1}^{n} p_i \frac{\partial x_i}{\partial v_j} = \frac{\partial u}{\partial v_j}, \qquad j = 1, \ldots, n - 1 \tag{47}$$

(where, of course, the p_i are evaluated on S). On the other hand, (41) must hold on S, i.e.,

$$F(x_1, \ldots, x_n, u, p_1, \ldots, p_n) = 0 \tag{48}$$

if we replace $x_1, \ldots, x_n$, u by their expressions in terms of $v_1, \ldots, v_{n-1}$. Together (47) and (48) form a system of n equations in the unknown functions $p_1, \ldots, p_n$ of the arguments $v_1, \ldots, v_{n-1}$. The implicit function theorem can be applied to this system, provided that

$$\left| \begin{array}{cccc} P_1 & P_2 & \cdots & P_n \\ \dfrac{\partial x_1}{\partial v_1} & \dfrac{\partial x_2}{\partial v_1} & \cdots & \dfrac{\partial x_n}{\partial v_1} \\ \cdots\cdots\cdots\cdots\cdots\cdots \\ \dfrac{\partial x_1}{\partial v_{n-1}} & \dfrac{\partial x_2}{\partial v_{n-1}} & \cdots & \dfrac{\partial x_n}{\partial v_{n-1}} \end{array} \right| \neq 0 \tag{49}$$

for all the relevant values of $v_1, \ldots, v_{n-1}$, i.e., everywhere on S. Therefore, from now on we shall assume not only that $u(v_1, \ldots, v_{n-1})$ is specified on S, but also that *values of $p_1, \ldots, p_n$ satisfying* (47)–(49) *have been chosen on S*, where, according to the implicit function theorem, $p_1, \ldots, p_n$ are continuous and have continuous derivatives with respect to $v_1, \ldots, v_{n-1}$.

It should be noted that in general, because of the nonlinearity of equation (48), the values of p_i on S for a given u are not uniquely determined. However, if all the above conditions are satisfied, and if

$$p_i^{(1)}(A) = p_i^{(2)}(A), \qquad i = 1, \ldots, n$$

at any point A of S, then it follows from the implicit function theorem and the condition (49) that

$$p_i^{(1)} \equiv p_i^{(2)}, \qquad i = 1, \ldots, n.$$

Geometrically, the condition (49) means that at every point $(x_1^0, \ldots, x_n^0)$ of S, the projection onto the plane $(x_1, \ldots, x_n)$ of the characteristic curve passing through the point $(x_1^0, \ldots, x_n^0, u^0, p_1^0, \ldots, p_n^0)$, where $u^0, p_1^0, \ldots, p_n^0$ are determined from the initial conditions, is not tangent to S (verify this).

The uniqueness of the solution of the system (44)–(46) now implies the uniqueness of the solution of the Cauchy problem. In fact, since F is assumed to have continuous derivatives up to order 2 with respect to all its arguments, it follows from (42) that the right-hand sides of (44)–(46) have continuous first derivatives with respect to all their arguments, and this guarantees the uniqueness of the solution of the system (44)–(46).

Next we turn to the proof of the existence of a solution of the Cauchy problem in some neighborhood of the surface S, assuming that S and the function u specified on S have the two properties in italics on p. 216, and moreover that values of $p_1, \ldots, p_n$ satisfying (47)–(49) have been chosen on S. Just as in Secs. 60–62, we confine ourselves to the case of two independent variables. Then equation (41) takes the form

$$F(x, y, z, p, q) = 0, \tag{50}$$

where

$$p = \frac{\partial z}{\partial x}, \qquad q = \frac{\partial z}{\partial y},$$

while the system of equations for the characteristic curves [in the five-dimensional space of points (x, y, z, p, q)] becomes

$$\frac{dx}{ds} = \frac{P}{\sqrt{P^2 + Q^2}}, \qquad \frac{dy}{ds} = \frac{Q}{\sqrt{P^2 + Q^2}}, \tag{51}$$

$$\frac{dp}{ds} = -\frac{X + pZ}{\sqrt{P^2 + Q^2}}, \qquad \frac{dq}{ds} = -\frac{Y + qZ}{\sqrt{P^2 + Q^2}}, \tag{52}$$

$$\frac{dz}{ds} = \frac{pP + qQ}{\sqrt{P^2 + Q^2}}, \tag{53}$$

where

$$X = \frac{\partial F}{\partial x}, \quad Y = \frac{\partial F}{\partial y}, \quad Z = \frac{\partial F}{\partial z}, \quad P = \frac{\partial F}{\partial p}, \quad Q = \frac{\partial F}{\partial q}.$$

The initial conditions are determined by the relations

$$x = x(v), \quad y = y(v), \quad z = z(v), \quad p = p(v), \quad q = q(v),$$

where the first three functions are twice continuously differentiable and

$$\left(\frac{dx}{dv}\right)^2 + \left(\frac{dy}{dv}\right)^2 > 0,$$

while the last two functions are continuously differentiable, and satisfy equation (50) and the conditions

$$p \frac{dx}{dv} + q \frac{dy}{dv} = \frac{dz}{dv} \tag{54}$$

$$\begin{vmatrix} P & Q \\ \dfrac{dx}{dv} & \dfrac{dy}{dv} \end{vmatrix} \neq 0. \tag{55}$$

To prove the existence of a solution under these conditions, we need only verify that the solutions[19]

$$x(s, v), \quad y(s, v), \quad z(s, v), \quad p(s, v), \quad q(s, v) \tag{56}$$

constructed from the system (51)–(53) and the initial data satisfies the following conditions:

1) The system of equations $x = x(s, v)$, $y = y(s, v)$ can be uniquely solved for s and v in some neighborhood $\mathcal{N}$ of the curve S, and the solution has continuous derivatives with respect to x and y. Then s and v can be chosen as curvilinear coordinates in $\mathcal{N}$, and if we replace s and v in $z(s, v)$ by their expressions in terms of x and y, we obtain z as a continuously differentiable function of x and y.[20]

2) The functions (56) satisfy equation (50) everywhere in $\mathcal{N}$.

3) The identities

$$p \equiv \frac{\partial z}{\partial x}, \qquad q \equiv \frac{\partial z}{\partial y} \tag{57}$$

   hold everywhere in $\mathcal{N}$.

---

[19] It is assumed that $s = 0$ on S.

[20] By hypothesis, the functions $x(v), \ldots, q(v)$ specified on S have continuous derivatives with respect to v, and the right-hand sides of (51)–(53) have continuous derivatives with respect to all their arguments. Therefore the functions (56) solving the system (51)–(53) have continuous derivatives with respect to s and v.

To prove condition 1, we need only show that the determinant

$$
\begin{vmatrix}
\dfrac{\partial x}{\partial s} & \dfrac{\partial y}{\partial s} \\[2ex]
\dfrac{\partial x}{\partial v} & \dfrac{\partial y}{\partial v}
\end{vmatrix}
\tag{58}
$$

is nonzero for all sufficiently small $|s|$. Since the elements of (58) are continuous in s and v, it is enough to show that (58) is nonzero on S itself. But this follows from (55) and (51).

As for condition 2, it obviously holds on the curve S, since the initial values of p and q have been chosen to satisfy (50) on S. To prove that (50) holds not only on the curve S (i.e., for $s = 0$), but also for all sufficiently small $|s|$, we show that substitution of the solution (56) of the system (51)–(53) for the quantities x, y, z, p, q appearing in the left-hand side of (50) leads to an expression which does not depend on s. In fact,

$$
\frac{dF}{ds} = X\frac{dx}{ds} + Y\frac{dy}{ds} + Z\frac{dz}{ds} + P\frac{dp}{ds} + Q\frac{dq}{ds},
$$

and replacing $dx/ds, \ldots, dq/ds$ by the right-hand sides of (51)–(53), we find that dF/ds equals zero.

To verify condition 3, we proceed indirectly, first proving that the two equations

$$
\frac{\partial z}{\partial v} - p\frac{\partial x}{\partial v} - q\frac{\partial y}{\partial v} = 0,
\tag{59}
$$

$$
\frac{\partial z}{\partial s} - p\frac{\partial x}{\partial s} - q\frac{\partial y}{\partial s} = 0.
\tag{60}
$$

Since, according to (51),

$$
\frac{\partial x}{\partial s} = \frac{P}{\sqrt{P^2 + Q^2}}, \qquad \frac{\partial y}{\partial s} = \frac{Q}{\sqrt{P^2 + Q^2}},
$$

equation (60) coincides with (53). As for equation (59), so far we only know that it holds for $s = 0$, since the initial values of p and q on the curve S have been chosen to satisfy (54). To prove that (59) is satisfied for nonzero s, we write

$$
U \equiv \frac{\partial z}{\partial v} - p\frac{\partial x}{\partial v} - q\frac{\partial y}{\partial v}
$$

and then calculate dU/ds. The differentiation with respect to s is easily justified. In fact, as already noted, the solution (56) of the system (51)–(53) has continuous derivatives with respect to s and v. Therefore, if (56) is substituted into (51)–(53), the right-hand sides of the resulting identities will have continuous derivatives with respect to s and v, and hence so will the left-hand sides, i.e., the derivatives

$$\frac{\partial^2 x}{\partial s\,\partial v}, \qquad \frac{\partial^2 y}{\partial s\,\partial v}, \qquad \frac{\partial^2 z}{\partial s\,\partial v}$$

exist and are continuous. It follows that

$$\frac{dU}{ds} = \frac{\partial^2 z}{\partial v\,\partial s} - \frac{\partial p}{\partial s}\frac{\partial x}{\partial v} - \frac{\partial q}{\partial s}\frac{\partial y}{\partial v} - p\frac{\partial^2 x}{\partial v\,\partial s} - q\frac{\partial^2 y}{\partial v\,\partial s}. \tag{61}$$

We now differentiate the identity

$$\frac{\partial z}{\partial s} - p\frac{\partial x}{\partial s} - q\frac{\partial y}{\partial s} \equiv 0$$

(valid everywhere in $\mathcal{N}$ by the argument just given) with respect to v, obtaining

$$\frac{\partial^2 z}{\partial s\,\partial v} - \frac{\partial p}{\partial v}\frac{\partial x}{\partial s} - \frac{\partial q}{\partial v}\frac{\partial y}{\partial s} - p\frac{\partial^2 x}{\partial s\,\partial v} - q\frac{\partial^2 y}{\partial s\,\partial v} \equiv 0. \tag{62}$$

Subtraction of (62) from (61) gives

$$\frac{dU}{ds} = \frac{\partial p}{\partial v}\frac{\partial x}{\partial s} + \frac{\partial q}{\partial v}\frac{\partial y}{\partial s} - \frac{\partial p}{\partial s}\frac{\partial x}{\partial v} - \frac{\partial q}{\partial s}\frac{\partial y}{\partial v},$$

or

$$\frac{dU}{ds} = \frac{\partial p}{\partial v}\frac{P}{\sqrt{P^2 + Q^2}} + \frac{\partial q}{\partial v}\frac{Q}{\sqrt{P^2 + Q^2}} + \frac{\partial x}{\partial v}\frac{X + pZ}{\sqrt{P^2 + Q^2}} + \frac{\partial y}{\partial v}\frac{Y + qZ}{\sqrt{P^2 + Q^2}} \tag{63}$$

after using (51) and (52). Moreover, differentiating the identity

$$F(x, y, z, p, q) \equiv 0$$

with respect to v, we obtain

$$X\frac{\partial x}{\partial v} + Y\frac{\partial y}{\partial v} + Z\frac{\partial z}{\partial v} + P\frac{\partial p}{\partial v} + Q\frac{\partial q}{\partial v} = 0. \tag{64}$$

Dividing both sides of (64) by $\sqrt{P^2 + Q^2}$ and subtracting the result from (63), we find that

$$\frac{dU}{ds} = -\frac{Z}{\sqrt{P^2 + Q^2}}\left(\frac{\partial z}{\partial v} - p\frac{\partial x}{\partial v} - q\frac{\partial y}{\partial v}\right) = -\frac{ZU}{\sqrt{P^2 + Q^2}},$$

and hence

$$U(s) = U(0)\exp\left\{-\int_0^s \frac{Z}{\sqrt{P^2 + Q^2}}\,ds\right\}.$$

Since $U(0) = 0$, it follows that $U(s) = 0$ for the other values of s.

Thus, finally, we have proved that the identities

$$\frac{\partial z}{\partial v} \equiv p \frac{\partial x}{\partial v} + q \frac{\partial y}{\partial v}, \qquad \frac{\partial z}{\partial s} \equiv p \frac{\partial x}{\partial s} + q \frac{\partial y}{\partial s} \tag{65}$$

hold in the whole neighborhood $\mathcal{N}$. Noting that

$$\frac{\partial z}{\partial x} = \frac{\partial z}{\partial v} \frac{\partial v}{\partial x} + \frac{\partial z}{\partial s} \frac{\partial s}{\partial x}, \qquad \frac{\partial z}{\partial y} = \frac{\partial z}{\partial v} \frac{\partial v}{\partial y} + \frac{\partial z}{\partial s} \frac{\partial s}{\partial y}, \tag{66}$$

and substituting (65) into (66), we obtain

$$\frac{\partial z}{\partial x} \equiv \left(p \frac{\partial x}{\partial v} + q \frac{\partial y}{\partial v} \right)\frac{\partial v}{\partial x} + \left(p \frac{\partial x}{\partial s} + q \frac{\partial y}{\partial s} \right)\frac{\partial s}{\partial x} = p \frac{\partial x}{\partial x} + q \frac{\partial y}{\partial x} = p,$$

and similarly

$$\frac{\partial z}{\partial y} \equiv q.$$

Therefore condition 3, p. 218 is satisfied, and the proof of the existence of a solution of the Cauchy problem is complete.

Remark. Instead of using s, the projection of the length of the characteristic curve onto the plane $(x_1, \ldots, x_n)$ as the parameter, we can introduce a new parameter t, related to s by the formula

$$ds = \sqrt{\sum_{j=1}^{n} P_j^2}\, dt.$$

Then the system (44)–(46) can be written in the form

$$\frac{dx_i}{dt} = P_i, \qquad i = 1, \ldots, n,$$

and so on.

Example. *Find the solution of the equation*

$$\left(\frac{\partial z}{\partial x}\right)^2 + \left(\frac{\partial z}{\partial y}\right)^2 - 1 = 0 \tag{67}$$

passing through the circle
$$x^2 + y^2 = 1, \qquad z = 0. \tag{68}$$

Introducing a parameter v, we write (68) in the form

$$x = \cos v, \quad y = \sin v, \quad z = 0. \tag{69}$$

In this case, equations (51)–(53) become

$$\frac{dx}{2p} = \frac{dy}{2q} = \frac{dz}{2(p^2 + q^2)} = \frac{dp}{0} = \frac{dq}{0} = dt, \tag{70}$$

and the last two equations imply $p = C_1$, $q = C_2$, where C_1 and C_2 are constants. Substituting these values into the first three equations, we obtain

$$x = 2C_3 t + C, \quad y = 2C_4 t + C, \quad z = 2(C_1^2 + C_2^2)t + C_5,$$

where C_3, C_4 and C_5 are constants. To satisfy (67), we must have

$$C_1^2 + C_2^2 = 1, \tag{71}$$

and hence

$$z = 2t + C_5.$$

Moreover,

$$C_3 = \cos v, \quad C_4 = \sin v, \quad C_5 = 0,$$

since the curve

$$x = 2C_1 t + C_3, \quad y = 2C_2 t + C_4, \quad z = 2(C_1^2 + C_2^2)t + C_5$$

must go through the point determined by the parameter v on the circle (69). Therefore the integral surface of equation (67) passing through the circle (69) is given by

$$x = 2C_1 t + \cos v, \quad y = 2C_2 t + \sin v, \quad z = 2t, \tag{72}$$

where t and v are parameters.

To satisfy the condition

$$\frac{\partial z}{\partial v} = p \frac{\partial x}{\partial v} + q \frac{\partial y}{\partial v}$$

for $t = 0$, it is necessary that

$$-p \sin v + q \cos v = 0$$

or

$$C_1 \sin v = C_2 \cos v,$$

and hence

$$C_1 = \varepsilon \cos v, \quad C_2 = \varepsilon \sin v, \quad \text{where } \varepsilon = \pm 1,$$

because of (71). Continuity considerations show that ε must have the same value along any characteristic curve, and hence the equations (72) take the form

$$x = (2t\varepsilon + 1) \cos v, \quad y = (2t\varepsilon + 1) \sin v, \quad z = 2t.$$

Elimination of t and v gives

$$x^2 + y^2 = (1 \pm z)^2. \tag{73}$$

Thus we have finally found two integral curves of equation (67) passing through the circle (68). These are two circular cones in (x, y, z) space with the circle (68) or (69) as base and the z-axis as common axis.

Remark. The above example shows that the requirement that the determinant (58) be nonzero only in the immediate vicinity of the curve S is an essential feature of the problem, since

$$\frac{\partial(x, y)}{\partial(t, v)}$$

vanishes for $t = \pm\frac{1}{2}$. Suppose the projections onto (x, y, z) space of the integral curves of the system (51)–(53) passing through the curve $\bar{S}$ (in the notation of Sec. 62) are given by

$$x = x(s, x^0, y^0), \quad y = y(s, x^0, y^0), \quad z = z(s, x^0, y^0). \tag{74}$$

Then it must not be thought that the curves (74) form a smooth surface for arbitrarily large values of the parameter s (or of the equivalent parameter t). For example, despite the fact that the integral curves of the system (70) are determined by the system for arbitrarily large t, the integral surfaces of equation (67) passing through the circle (69) cannot be continued indefinitely without eventually arriving at singular points, i.e., the vertices of the cones (73). Singularities of this type cannot occur in the case of quasi-linear equations, since the characteristic curves of such equations do not intersect in $(x_1, \ldots, x_n, u)$ space.

Problem 1. Clarify the relation between the characteristic curves introduced in this section and those of Secs. 60 and 62.

Problem 2. A characteristic curve might be thought of as a curve in $(x_1, \ldots, x_n, u)$ space through each point of which passes a plane whose normal has direction numbers $p_1, \ldots, p_n, -1$. Prove that this plane is tangent to the curve itself at each point of the curve.

Problem 3 (A. Haar). Let $\bar{G}$ be the pyramid in $(x_1, \ldots, x_n)$ space $(n \geqslant 2)$ defined by the inequalities

$$0 \leqslant x_n \leqslant a, \quad \max\{|x_1|, \ldots, |x_{n-1}|\} \leqslant a - x_n \quad (a > 0),$$

and let M be the base of $\bar{G}$. Suppose $u(x_1, \ldots, x_n)$ is a differentiable function defined on $\bar{G}$ satisfying the inequality

$$\left|\frac{\partial u}{\partial x_n}\right| \leqslant \sum_{i=1}^{n-1} \left|\frac{\partial u}{\partial x_i}\right| + |u| + \delta \quad (\delta \geqslant 0).$$

Prove that then

$$\min\left\{\min_M u, 0\right\} e^{x_n} - \delta(e^{x_n} - 1) \leqslant u(x_1, \ldots, x_n)$$
$$\leqslant \max\left\{\max_M u, 0\right\} e^{x_n} + \delta(e^{x_n} - 1)$$

everywhere in $\bar{G}$.

Hint. Consider the point where the function

$$v(x_1, \ldots, x_n; \lambda) = e^{-(1+\lambda)x_n}[u(x_1, \ldots, x_n) - \delta(e^{x_n} - 1)] \quad (\lambda > 0)$$

takes its maximum.

Problem 4. Use Prob. 3 and Hadamard's lemma to prove (under suitable conditions) that

a) The solution of equation (41) is unique;

b) The solution of (41) depends continuously on the initial data and the form of the equation.

Hint. For simplicity, assume first that the initial conditions are specified for $x_n = 0$.

65. Pfaffian Equations

An equation of the form

$$P \, dx + Q \, dy + R \, dz = 0, \tag{75}$$

where P, Q and R are functions of x, y and z, is called a *Pfaffian equation* (in three dimensions). There are two approaches to the study of such equations. In the first approach, x, y and z are regarded as functions of a single parameter t. Then specifying two of the variables x, y and z as functions of t, we arrive at an ordinary differential equation determining the third variable. We can also require x, y and z to obey an arbitrary relation of the form

$$\Phi(x, y, z) = 0. \tag{76}$$

Then, regarding x, y and z in (76) as functions of a parameter t and differentiating with respect to t, we obtain

$$\frac{\partial \Phi}{\partial x} \, dx + \frac{\partial \Phi}{\partial y} \, dy + \frac{\partial \Phi}{\partial z} \, dz = 0. \tag{77}$$

Under very general assumptions concerning Φ, P, Q and R, equations (75) and (77) can be solved for the ratios of any two differentials to the third, e.g., for dy/dx and dz/dx. This gives a system of two ordinary differential equations determining y and z as functions of x. In general, the condition (76) leaves one arbitrary constant in the general solution.

In the second approach to Pfaffian equations, one of the variables x, y and z, say z, is regarded as a function of the other two. Suppose $R \neq 0$ in the domain under consideration. Then (75) implies

$$dz = P_1 \, dx + Q_1 \, dy, \tag{78}$$

where

$$P_1 = -\frac{P}{R}, \qquad Q_1 = -\frac{Q}{R},$$

and hence

$$\frac{\partial z}{\partial x} = P_1(x, y, z), \tag{78'}$$

$$\frac{\partial z}{\partial y} = Q_1(x, y, z). \tag{78''}$$

Suppose it is known in advance that z has continuous second derivatives with respect to x and y, while P_1 and Q_1 have continuous first derivatives with respect to all their arguments. Then

$$\frac{\partial P_1}{\partial y} + \frac{\partial P_1}{\partial z}\frac{\partial z}{\partial y} = \frac{\partial Q_1}{\partial x} + \frac{\partial Q_1}{\partial z}\frac{\partial z}{\partial x}$$

or

$$\frac{\partial P_1}{\partial y} + \frac{\partial P_1}{\partial z}Q_1 = \frac{\partial Q_1}{\partial x} + \frac{\partial Q_1}{\partial z}P_1, \tag{79}$$

since

$$\frac{\partial^2 z}{\partial x\,\partial y} = \frac{\partial^2 z}{\partial y\,\partial x}.$$

In differentiating P_1 and Q_1 with respect to x and y, we have taken into account not only the explicit dependence of P_1 and Q_1 on x and y, but also their implicit dependence on x and y via the variable z (which, by hypothesis, is a function of x and y). The condition (79) can also be written in the form

$$P\left(\frac{\partial Q}{\partial z} - \frac{\partial P}{\partial y}\right) + Q\left(\frac{\partial R}{\partial x} - \frac{\partial P}{\partial z}\right) + R\left(\frac{\partial P}{\partial y} - \frac{\partial Q}{\partial x}\right) = 0. \tag{80}$$

THEOREM. *Suppose the condition* (80), *or equivalently* (79), *is satisfied identically on a three-dimensional domain G, where P_1 and Q_1 have continuous derivatives up to order 2 (inclusive) with respect to all their arguments. Then one and only one integral surface of equation* (78), *or equivalently of* (75), *passes through each point (x^0, y^0, z^0) of G.*

Proof. If we set $y = y^0$ in (78′), the resulting equation determines a unique integral curve L in the plane $y = y^0$ passing through the point (x^0, y^0, z^0). Similarly, equation (78″) with x assigned a constant value C determines a unique integral curve l_C in the plane $x = C$ passing through the point of L lying in this plane. Then the set of curves l_C constructed in this way for all points of L uniquely determines a surface S passing through the point (x^0, y^0, z^0). If the surface S has equation $z = z(x, y)$, then $z(x, y)$ has continuous derivatives up to order 2, because of the continuous dependence of the solution of (78″) on the parameter C and on the initial data.

Next we show that S is actually an integral surface of (78). The fact that $z(x, y)$ satisfies (78″) is obvious from the construction of S, and similarly $z(x, y)$ satisfies (78′) for $y = y^0$. To show that $z(x, y)$ satisfies (78′) for other values of y, we consider the function

$$F = \frac{\partial z}{\partial x} - P_1(x, y, z)$$

and evaluate $\partial F/\partial y$, obtaining

$$
\frac{\partial F}{\partial y} = \frac{\partial}{\partial y}\left(\frac{\partial z}{\partial x}\right) - \frac{\partial P_1}{\partial y} - \frac{\partial P_1}{\partial z}\frac{\partial z}{\partial y} = \frac{\partial Q_1}{\partial x} + \frac{\partial Q_1}{\partial z}\frac{\partial z}{\partial x} - \frac{\partial P_1}{\partial y} - \frac{\partial P_1}{\partial z}\frac{\partial z}{\partial y}
$$

$$
= \frac{\partial Q_1}{\partial x} + \frac{\partial Q_1}{\partial z}P_1 + \frac{\partial Q_1}{\partial z}F - \frac{\partial P_1}{\partial y} - \frac{\partial P_1}{\partial z}Q_1,
$$

where the formulas

$$
\frac{\partial}{\partial y}\left(\frac{\partial z}{\partial x}\right) = \frac{\partial}{\partial x}\left(\frac{\partial z}{\partial y}\right), \qquad \frac{\partial z}{\partial y} = Q_1
$$

have been used. Substituting (72) into (74), we find that (74) reduces to

$$
\frac{\partial F}{\partial y} = \frac{\partial Q_1}{\partial z}F_1,
$$

and hence

$$
F(x, y) = F(x, y^0)\left\{\exp \int_{y^0}^{y} \frac{\partial Q_1}{\partial z}\, dy\right\}.
$$

Therefore $F(x, y)$ vanishes for all y under consideration, since it vanishes for $y = y^0$. This completes the proof.

Remark. Geometrically, the first way of solving Pfaffian equations corresponds to constructing curves orthogonal to the given three-dimensional direction field (with directions specified by the vector with components P, Q and R). The second approach corresponds to constructing surfaces orthogonal to the same direction field (or equivalently, surfaces with given tangent planes at every point of space).

Problem. Consider the n-dimensional Pfaffian equation

$$
\sum_{i=1}^{n} a_i(x_1, \ldots, x_n)\, dx_i = 0, \tag{81}
$$

with appropriate assumptions concerning the coefficients a_i. In how many different ways can this equation be treated? Analyze in most detail the approach in which the "integral manifolds" are $(n - 1)$-dimensional surfaces.

Hint. It is convenient to write (81) as a system

$$
\frac{\partial u}{\partial x_i} = b_i(x_1, \ldots, x_{n-1}, u), \qquad i = 1, \ldots, n - 1,
$$

where $u = x_n$.

BIBLIOGRAPHY[1]

Birkhoff, G. and G.-C. Rota, *Ordinary Differential Equations*, Ginn and Co., Boston (1962).

Coddington, E. A., *An Introduction to Ordinary Differential Equations*, Prentice-Hall, Inc., Englewood Cliffs, N.J. (1961).

Coddington, E. A. and N. Levinson, *Theory of Ordinary Differential Equations*, McGraw-Hill Book Co., New York (1955).

Greenspan, D., *Theory and Solution of Ordinary Differential Equations*, The Macmillan Co., New York (1960).

Hartman, P., *Ordinary Differential Equations*, John Wiley and Sons, Inc., New York (1964).

Hochstadt, H., *Differential Equations, A Modern Approach*, Holt, Rinehart and Winston, New York (1964).

Hurewicz, W., *Lectures on Ordinary Differential Equations*, M.I.T. Technology Press, Cambridge, Mass. (1958).

Ince, E. L., *Ordinary Differential Equations*, Dover Publications, Inc., New York (1956).

Kamke, E., *Differentialgleichungen, Lösungsmethoden und Lösungen, I. Gewöhnliche Differentialgleichungen*, fifth edition, Akademische Verlagsgesellschaft, Geest & Portig K.-G., Leipzig (1956).

Pontryagin, L. S., *Ordinary Differential Equations* (translated by L. Kacinskas and W. B. Counts), Addison-Wesley Publishing Co., Inc., Reading, Mass. (1962).

---

[1] See also the books cited in Secs. 24, 38, and 63.

INDEX

A CATALOGUE OF SELECTED
DOVER SCIENCE BOOKS

A CATALOGUE OF SELECTED
DOVER SCIENCE BOOKS

Physics: The Pioneer Science, Lloyd W. Taylor. Very thorough non-mathematical survey of physics in a historical framework which shows development of ideas. Easily followed by laymen; used in dozens of schools and colleges for survey courses. Richly illustrated. Volume 1: Heat, sound, mechanics. Volume 2: Light, electricity. Total of 763 illustrations. Total of cvi + 847pp.

60565-5, 60566-3 Two volumes, Paperbound 5.50

THE RISE OF THE NEW PHYSICS, A. d'Abro. Most thorough explanation in print of central core of mathematical physics, both classical and modern, from Newton to Dirac and Heisenberg. Both history and exposition: philosophy of science, causality, explanations of higher mathematics, analytical mechanics, electromagnetism, thermodynamics, phase rule, special and general relativity, matrices. No higher mathematics needed to follow exposition, though treatment is elementary to intermediate in level. Recommended to serious student who wishes verbal understanding. 97 illustrations. Total of ix + 982pp.

20003-5, 20004-3 Two volumes, Paperbound $6.00

INTRODUCTION TO CHEMICAL PHYSICS, John C. Slater. A work intended to bridge the gap between chemistry and physics. Text divided into three parts: Thermodynamics, Statistical Mechanics, and Kinetic Theory; Gases, Liquids and Solids; and Atoms, Molecules and the Structure of Matter, which form the basis of the approach. Level is advanced undergraduate to graduate, but theoretical physics held to minimum. 40 tables, 118 figures. xiv + 522pp.

62562-1 Paperbound $4.00

BASIC THEORIES OF PHYSICS, Peter C. Bergmann. Critical examination of important topics in classical and modern physics. Exceptionally useful in examining conceptual framework and methodology used in construction of theory. Excellent supplement to any course, textbook. Relatively advanced.
Volume 1. Heat and Quanta. Kinetic hypothesis, physics and statistics, stationary ensembles, thermodynamics, early quantum theories, atomic spectra, probability waves, quantization in wave mechanics, approximation methods, abstract quantum theory. 8 figures. x + 300pp.   60968-5 Paperbound $2.50
Volume 2. Mechanics and Electrodynamics. Classical mechanics, electro- and magnetostatics, electromagnetic induction, field waves, special relativity, waves, etc. 16 figures, viii + 260pp.   60969-3 Paperbound $2.75

FOUNDATIONS OF PHYSICS, Robert Bruce Lindsay and Henry Margenau. Methods and concepts at the heart of physics (space and time, mechanics, probability, statistics, relativity, quantum theory) explained in a text that bridges gap between semi-popular and rigorous introductions. Elementary calculus assumed. "Thorough and yet not over-detailed," *Nature*. 35 figures. xviii + 537 pp.

60377-6 Paperbound $3.50

AN ELEMENTARY INTRODUCTION TO THE THEORY OF PROBABILITY, B. V. Gnedenko and A. Ya. Khinchin. Introduction to facts and principles of probability theory. Extremely thorough within its range. Mathematics employed held to elementary level. Excellent, highly accurate layman's introduction. Translated from the fifth Russian edition by Leo Y. Boron. xii + 130pp.
60155-2 Paperbound $2.00

SELECTED PAPERS ON NOISE AND STOCHASTIC PROCESSES, edited by Nelson Wax. Six papers which serve as an introduction to advanced noise theory and fluctuation phenomena, or as a reference tool for electrical engineers whose work involves noise characteristics, Brownian motion, statistical mechanics. Papers are by Chandrasekhar, Doob, Kac, Ming, Ornstein, Rice, and Uhlenbeck. Exact facsimile of the papers as they appeared in scientific journals. 19 figures. v + 337pp. 6⅛ x 9¼.
60262-1 Paperbound $3.50

STATISTICS MANUAL, Edwin L. Crow, Frances A. Davis and Margaret W. Maxfield. Comprehensive, practical collection of classical and modern methods of making statistical inferences, prepared by U. S. Naval Ordnance Test Station. Formulae, explanations, methods of application are given, with stress on use. Basic knowledge of statistics is assumed. 21 tables, 11 charts, 95 illustrations. xvii + 288pp.
60599-X Paperbound $2.50

MATHEMATICAL FOUNDATIONS OF INFORMATION THEORY, A. I. Khinchin. Comprehensive introduction to work of Shannon, McMillan, Feinstein and Khinchin, placing these investigations on a rigorous mathematical basis. Covers entropy concept in probability theory, uniqueness theorem, Shannon's inequality, ergodic sources, the E property, martingale concept, noise, Feinstein's fundamental lemma, Shanon's first and second theorems. Translated by R. A. Silverman and M. D. Friedman. iii + 120pp.
60434-9 Paperbound $1.75

INTRODUCTION TO SYMBOLIC LOGIC AND ITS APPLICATION, Rudolf Carnap. Clear, comprehensive, rigorous introduction. Analysis of several logical languages. Investigation of applications to physics, mathematics, similar areas. Translated by Wiliam H. Meyer and John Wilkinson. xiv + 214pp.
60453-5 Paperbound $2.50

SYMBOLIC LOGIC, Clarence I. Lewis and Cooper H. Langtord. Probably the most cited book in the literature, with much material not otherwise obtainable. Paradoxes, logic of extensions and intensions, converse substitution, matrix system, strict limitations, existence of terms, truth value systems, similar material. vii + 518pp.
60170-6 Paperbound $4.50

VECTOR AND TENSOR ANALYSIS, George E. Hay. Clear introduction; starts with simple definitions, finishes with mastery of oriented Cartesian vectors, Christoffel symbols, solenoidal tensors, and applications. Many worked problems show applications. 66 figures. viii + 193pp.
60109-9 Paperbound $2.50

THE ELEMENTS OF NON-EUCLIDEAN GEOMETRY, Duncan M. Y. Sommerville. Presentation of the development of non-Euclidean geometry in logical order, from a fundamental analysis of the concept of parallelism to such advanced topics as inversion, transformations, pseudosphere, geodesic representation, relation between parataxy and parallelism, etc. Knowledge of only high-school algebra and geometry is presupposed. 126 problems, 129 figures. xvi + 274pp.
60460-8 Paperbound $2.50

NON-EUCLIDEAN GEOMETRY: A CRITICAL AND HISTORICAL STUDY OF ITS DEVELOPMENT, Roberto Bonola. Standard survey, clear, penetrating, discussing many systems not usually represented in general studies. Easily followed by non-specialist. Translated by H. Carslaw. Bound in are two most important texts: Bolyai's "The Science of Absolute Space" and Lobachevski's "The Theory of Parallels," translated by G. B. Halsted. Introduction by F. Enriques. 181 diagrams. Total of 431pp.
60027-0 Paperbound $3.00

ELEMENTS OF NUMBER THEORY, Ivan M. Vinogradov. By stressing demonstrations and problems, this modern text can be understood by students without advanced math backgrounds. "A very welcome addition," *Bulletin, American Mathematical Society.* Translated by Saul Kravetz. Over 200 fully-worked problems. 100 numerical exercises. viii + 227pp.
60259-1 Paperbound $2.50

THEORY OF SETS, E. Kamke. Lucid introduction to theory of sets, surveying discoveries of Cantor, Russell, Weierstrass, Zermelo, Bernstein, Dedekind, etc. Knowledge of college algebra is sufficient background. "Exceptionally well written," *School Science and Mathematics.* Translated by Frederick Bagemihl. vii + 144pp.
60141-2 Paperbound $1.75

A TREATISE ON THE DIFFERENTIAL GEOMETRY OF CURVES AND SURFACES, Luther P. Eisenhart. Detailed, concrete introductory treatise on differential geometry, developed from author's graduate courses at Princeton University. Thorough explanation of the geometry of curves and surfaces, concentrating on problems most helpful to students. 683 problems, 30 diagrams. xiv + 474pp.
60667-8 Paperbound $3.50

AN ESSAY ON THE FOUNDATIONS OF GEOMETRY, Bertrand Russell. A mathematical and physical analysis of the place of the a priori in geometric knowledge. Includes critical review of 19th-century work in non-Euclidean geometry as well as illuminating insights of one of the great minds of our time. New foreword by Morris Kline. xx + 201pp.
60233-8 Paperbound $2.50

INTRODUCTION TO THE THEORY OF NUMBERS, Leonard E. Dickson. Thorough, comprehensive approach with adequate coverage of classical literature, yet simple enough for beginners. Divisibility, congruences, quadratic residues, binary quadratic forms, primes, least residues, Fermat's theorem, Gauss's lemma, and other important topics. 249 problems, 1 figure. viii + 183pp.
60342-3 Paperbound $2.00

CONTRIBUTIONS TO THE FOUNDING OF THE THEORY OF TRANSFINITE NUMBERS, Georg Cantor. The famous articles of 1895-1897 which founded a new branch of mathematics, translated with 82-page introduction by P. Jourdain. Not only a great classic but still one of the best introductions for the student. ix + 211pp.
60045-9 Paperbound $2.50

ESSAYS ON THE THEORY OF NUMBERS, Richard Dedekind. Two classic essays, on the theory of irrationals, giving an arithmetic and rigorous foundation; and on transfinite numbers and properties of natural numbers. Translated by W. W. Beman. iii + 115pp.
21010-3 Paperbound $1.75

GEOMETRY OF FOUR DIMENSIONS, H. P. Manning. Part verbal, part mathematical development of fourth dimensional geometry. Historical introduction. Detailed treatment is by synthetic method, approaching subject through Euclidean geometry. No knowledge of higher mathematics necessary. 76 figures. ix + 348pp.
60182-X Paperbound $3.00

AN INTRODUCTION TO THE GEOMETRY OF N DIMENSIONS, Duncan M. Y. Sommerville. The only work in English devoted to higher-dimensional geometry. Both metric and projective properties of n-dimensional geometry are covered. Covers fundamental ideas of incidence, parallelism, perpendicularity, angles between linear space, enumerative geometry, analytical geometry, polytopes, analysis situs, hyperspacial figures. 60 diagrams. xvii + 196pp.
60494-2 Paperbound $2.00

THE THEORY OF SOUND, J. W. S. Rayleigh. Still valuable classic by the great Nobel Laureate. Standard compendium summing up previous research and Rayleigh's original contributions. Covers harmonic vibrations, vibrating systems, vibrations of strings, membranes, plates, curved shells, tubes, solid bodies, refraction of plane waves, general equations. New historical introduction and bibliography by R. B. Lindsay, Brown University. 97 figures. lviii + 984pp.
60292-3, 60293-1 Two volumes, Paperbound $6.00

ELECTROMAGNETIC THEORY: A CRITICAL EXAMINATION OF FUNDAMENTALS, Alfred O'Rahilly. Critical analysis and restructuring of the basic theories and ideas of classical electromagnetics. Analysis is carried out through study of the primary treatises of Maxwell, Lorentz, Einstein, Weyl, etc., which established the theory. Expansive reference to and direct quotation from these treatises. Formerly *Electromagnetics*. Total of xvii + 884pp.
60126-9, 60127-7 Two volumes, Paperbound $6.00

ELEMENTARY CONCEPTS OF TOPOLOGY, Paul Alexandroff. Elegant, intuitive approach to topology, from the basic concepts of set-theoretic topology to the concept of Betti groups. Stresses concepts of complex, cycle and homology. Shows how concepts of topology are useful in math and physics. Introduction by David Hilbert. Translated by Alan E. Farley. 25 figures. iv + 57pp.
60747-X Paperbound $1.25

MICROSCOPY FOR CHEMISTS, Harold F. Schaeffer. Thorough text; operation of microscope, optics, photomicrographs, hot stage, polarized light, chemical procedures for organic and inorganic reactions. 32 specific experiments cover specific analyses: industrial, metals, other important subjects. 136 figures. 264pp.
61682-7 Paperbound $2.50

OPTICKS, Sir Isaac Newton. A survey of 18th-century knowledge on all aspects of light as well as a description of Newton's experiments with spectroscopy, colors, lenses, reflection, refraction, theory of waves, etc. in language the layman can follow. Foreword by Albert Einstein. Introduction by Sir Edmund Whittaker. Preface by I. Bernard Cohen. cxxvi + 406pp.   60205-2 Paperbound $4.00

LIGHT: PRINCIPLES AND EXPERIMENTS, George S. Monk. Thorough coverage, for student with background in physics and math, of physical and geometric optics. Also includes 23 experiments on optical systems, instruments, etc. "Probably the best intermediate text on optics in the English language," *Physics Forum.* 275 figures. xi + 489pp.   60341-5 Paperbound $3.50

PHYSICAL OPTICS, Robert W. Wood. A classic in the field, this is a valuable source for students of physical optics and excellent background material for a study of electromagnetic theory. Partial contents: nature and rectilinear propagation of light, reflection from plane and curved surfaces, refraction, absorption and dispersion, origin of spectra, interference, diffraction, polarization, Raman effect, optical properties of metals, resonance radiation and fluorescence of atoms, magneto-optics, electro-optics, thermal radiation. 462 diagrams, 17 plates. xvi + 846pp.
61808-0 Paperbound $4.50

MIRRORS, PRISMS AND LENSES: A TEXTBOOK OF GEOMETRICAL OPTICS, James P. C. Southall. Introductory-level account of modern optical instrument theory, covering unusually wide range: lights and shadows, reflection of light and plane mirrors, refraction, astigmatic lenses, compound systems, aperture and field of optical system, the eye, dispersion and achromatism, rays of finite slope, the microscope, much more. Strong emphasis on earlier, elementary portions of field, utilizing simplest mathematics wherever possible. Problems. 329 figures. xxiv + 806pp.   61234-1 Paperbound $5.00

THE PSYCHOLOGY OF INVENTION IN THE MATHEMATICAL FIELD, Jacques Hadamard. Important French mathematician examines psychological origin of ideas, role of the unconscious, importance of visualization, etc. Based on own experiences and reports by Dalton, Pascal, Descartes, Einstein, Poincaré, Helmholtz, etc. xiii + 145pp.   20107-4 Paperbound $1.50

INTRODUCTION TO CHEMICAL PHYSICS, John C. Slater. A work intended to bridge the gap between chemistry and physics. Text divided into three parts: Thermodynamics, Statistical Mechanics, and Kinetic Theory; Gases, Liquids and Solids; and Atoms, Molecules and the Structure of Matter, which form the basis of the approach. Level is advanced undergraduate to graduate, but theoretical physics held to minimum. 40 tables, 118 figures. xiv + 522pp.
62562-1 Paperbound $4.00

MATHEMATICAL FOUNDATIONS OF STATISTICAL MECHANICS, A. I. Khinchin. Introduction to modern statistical mechanics: phase space, ergodic problems, theory of probability, central limit theorem, ideal monatomic gas, foundation of thermodynamics, dispersion and distribution of sum functions. Provides mathematically rigorous treatment and excellent analytical tools. Translated by George Gamow. viii + 179pp.   60147-1 Paperbound $2.50

INTRODUCTION TO PHYSICAL STATISTICS, Robert B. Lindsay. Elementary probability theory, laws of thermodynamics, classical Maxwell-Boltzmann statistics, classical statistical mechanics, quantum mechanics, other areas of physics that can be studied statistically. Full coverage of methods; basic background theory. ix + 306pp.   61882-X Paperbound $2.75

DIALOGUES CONCERNING TWO NEW SCIENCES, Galileo Galilei. Written near the end of Galileo's life and encompassing 30 years of experiment and thought, these dialogues deal with geometric demonstrations of fracture of solid bodies, cohesion, leverage, speed of light and sound, pendulums, falling bodies, accelerated motion, etc. Translated by Henry Crew and Alfonso de Salvio. Introduction by Antonio Favaro. xxiii + 300pp.   60099-8 Paperbound $2.25

FOUNDATIONS OF SCIENCE: THE PHILOSOPHY OF THEORY AND EXPERIMENT, Norman R. Campbell. Fundamental concepts of science examined on middle level: acceptance of propositions and axioms, presuppositions of scientific thought, scientific law, multiplication of probabilities, nature of experiment, application of mathematics, measurement, numerical laws and theories, error, etc. Stress on physics, but holds for other sciences. "Unreservedly recommended," Nature (England). Formerly Physics: The Elements. ix + 565pp.   60372-5 Paperbound $4.00

THE PHASE RULE AND ITS APPLICATIONS, Alexander Findlay, A. N. Campbell and N. O. Smith. Findlay's well-known classic, updated (1951). Full standard text and thorough reference, particularly useful for graduate students. Covers chemical phenomena of one, two, three, four and multiple component systems. "Should rank as the standard work in English on the subject," Nature. 236 figures. xii + 494pp.   60091-2 Paperbound $3.50

THERMODYNAMICS, Enrico Fermi. A classic of modern science. Clear, organized treatment of systems, first and second laws, entropy, thermodynamic potentials, gaseous reactions, dilute solutions, entropy constant. No math beyond calculus is needed, but readers are assumed to be familiar with fundamentals of thermometry, calorimetry. 22 illustrations. 25 problems. x + 160pp.
60361-X Paperbound $2.00

TREATISE ON THERMODYNAMICS, Max Planck. Classic, still recognized as one of the best introductions to thermodynamics. Based on Planck's original papers, it presents a concise and logical view of the entire field, building physical and chemical laws from basic empirical facts. Planck considers fundamental definitions, first and second principles of thermodynamics, and applications to special states of equilibrium. Numerous worked examples. Translated by Alexander Ogg. 5 figures. xiv + 297pp.   60219-2 Paperbound $2.50

EINSTEIN'S THEORY OF RELATIVITY, Max Born. Relativity theory analyzed, explained for intelligent layman or student with some physical, mathematical background. Includes Lorentz, Minkowski, and others. Excellent verbal account for teachers. Generally considered the finest non-technical account. vii + 376pp.
60769-0 Paperbound $2.75

PHYSICAL PRINCIPLES OF THE QUANTUM THEORY, Werner Heisenberg. Nobel Laureate discusses quantum theory, uncertainty principle, wave mechanics, work of Dirac, Schroedinger, Compton, Wilson, Einstein, etc. Middle, non-mathematical level for physicist, chemist not specializing in quantum; mathematical appendix for specialists. Translated by C. Eckart and F. Hoyt. 19 figures. viii + 184pp.
60113-7 Paperbound $2.00

PRINCIPLES OF QUANTUM MECHANICS, William V. Houston. For student with working knowledge of elementary mathematical physics; uses Schroedinger's wave mechanics. Evidence for quantum theory, postulates of quantum mechanics, applications in spectroscopy, collision problems, electrons, similar topics. 21 figures. 288pp.
60524-8 Paperbound $3.00

ATOMIC SPECTRA AND ATOMIC STRUCTURE, Gerhard Herzberg. One of the best introductions to atomic spectra and their relationship to structure; especially suited to specialists in other fields who require a comprehensive basic knowledge. Treatment is physical rather than mathematical. 2nd edition. Translated by J. W. T. Spinks. 80 illustrations. xiv + 257pp.
60115-3 Paperbound $2.00

ATOMIC PHYSICS: AN ATOMIC DESCRIPTION OF PHYSICAL PHENOMENA, Gaylord P. Harnwell and William E. Stephens. One of the best introductions to modern quantum ideas. Emphasis on the extension of classical physics into the realms of atomic phenomena and the evolution of quantum concepts. 156 problems. 173 figures and tables. xi + 401pp.
61584-7 Paperbound $3.00

ATOMS, MOLECULES AND QUANTA, Arthur E. Ruark and Harold C. Urey. 1964 edition of work that has been a favorite of students and teachers for 30 years. Origins and major experimental data of quantum theory, development of concepts of atomic and molecular structure prior to new mechanics, laws and basic ideas of quantum mechanics, wave mechanics, matrix mechanics, general theory of quantum dynamics. Very thorough, lucid presentation for advanced students. 230 figures. Total of xxiii + 810pp.
61106-X, 61107-8 Two volumes, Paperbound $6.00

INVESTIGATIONS ON THE THEORY OF THE BROWNIAN MOVEMENT, Albert Einstein. Five papers (1905-1908) investigating the dynamics of Brownian motion and evolving an elementary theory of interest to mathematicians, chemists and physical scientists. Notes by R. Fürth, the editor, discuss the history of study of Brownian movement, elucidate the text and analyze the significance of the papers. Translated by A. D. Cowper. 3 figures. iv + 122pp.
60304-0 Paperbound $1.50

ASTRONOMY AND COSMOGONY, Sir James Jeans. Modern classic of exposition, Jean's latest work. Descriptive astronomy, atrophysics, stellar dynamics, cosmology, presented on intermediate level. 16 illustrations. Preface by Lloyd Motz. xv + 428pp.   60923-5 Paperbound $3.50

EXPERIMENTAL SPECTROSCOPY, Ralph A. Sawyer. Discussion of techniques and principles of prism and grating spectrographs used in research. Full treatment of apparatus, construction, mounting, photographic process, spectrochemical analysis, theory. Mathematics kept to a minimum. Revised (1961) edition. 110 illustrations. x + 358pp.   61045-4 Paperbound $3.50

THEORY OF FLIGHT, Richard von Mises. Introduction to fluid dynamics, explaining fully the physical phenomena and mathematical concepts of aeronautical engineering, general theory of stability, dynamics of incompressible fluids and wing theory. Still widely recommended for clarity, though limited to situations in which air compressibility effects are unimportant. New introduction by K. H. Hohenemser. 408 figures. xvi + 629pp.   60541-8 Paperbound $5.00

AIRPLANE STRUCTURAL ANALYSIS AND DESIGN, Ernest E. Sechler and Louis G. Dunn. Valuable source work to the aircraft and missile designer: applied and design loads, stress-strain, frame analysis, plates under normal pressure, engine mounts, landing gears, etc. 47 problems. 256 figures. xi + 420pp.
                                                          61043-8 Paperbound $3.50

PHOTOELASTICITY: PRINCIPLES AND METHODS, H. T. Jessop and F. C. Harris. An introduction to general and modern developments in 2- and 3-dimensional stress analysis techniques. More advanced mathematical treatment given in appendices. 164 figures. viii + 184pp. 6⅛ x 9¼.   (USO) 60720-8 Paperbound $2.50

THE MEASUREMENT OF POWER SPECTRA FROM THE POINT OF VIEW OF COMMUNICATIONS ENGINEERING, Ralph B. Blackman and John W. Tukey. Techniques for measuring the power spectrum using elementary transmission theory and theory of statistical estimation. Methods of acquiring sound data, procedures for reducing data to meaningful estimates, ways of interpreting estimates. 36 figures and tables. Index. x + 190pp.   60507-8 Paperbound $2.50

GASEOUS CONDUCTORS: THEORY AND ENGINEERING APPLICATIONS, James D. Cobine. An indispensable reference for radio engineers, physicists and lighting engineers. Physical backgrounds, theory of space charges, applications in circuit interrupters, rectifiers, oscillographs, etc. 83 problems. Over 600 figures. xx + 606pp.   60442-X Paperbound $3.75

Prices subject to change without notice.